ROUTLEDGE LIBRARY EDITIONS: CONSERVATION

Volume 8

# CAMPAIGNING FOR THE ENVIRONMENT

# CAMPAIGNING FOR THE ENVIRONMENT

Edited by
RICHARD KIMBER AND
J. J. RICHARDSON

Routledge
Taylor & Francis Group
LONDON AND NEW YORK

First published in 1974 by Routledge & Kegan Paul Ltd

This edition first published in 2020
by Routledge
2 Park Square, Milton Park, Abingdon, Oxon OX14 4RN

and by Routledge
52 Vanderbilt Avenue, New York, NY 10017

*Routledge is an imprint of the Taylor & Francis Group, an informa business*



*British Library Cataloguing in Publication Data*
A catalogue record for this book is available from the British Library

ISBN: 978-0-367-43303-1 (Set)
ISBN: 978-1-00-300237-6 (Set) (ebk)
ISBN: 978-0-367-40952-4 (Volume 8) (hbk)
ISBN: 978-0-367-81011-5 (Volume 8) (ebk)

**Publisher's Note**
The publisher has gone to great lengths to ensure the quality of this reprint but points out that some imperfections in the original copies may be apparent.

**Disclaimer**
The publisher has made every effort to trace copyright holders and would welcome correspondence from those they have been unable to trace.

# Campaigning for the environment

*edited by*

**Richard Kimber**
**and J. J. Richardson**
Department of Politics, University of Keele

Routledge & Kegan Paul
London and Boston

*First published in 1974*
*by Routledge & Kegan Paul Ltd*
*Broadway House, 68-74 Carter Lane,*
*London EC4V 5EL and*
*9 Park Street,*
*Boston, Mass. 02108, USA*
*Set in Monotype Plantin*
*and printed in Great Britain at*
*The St Ann's Press, Park Road, Altrincham,*
*Cheshire WA14 5QQ*


*ISBN* 0 7100 7851

*Library of Congress Catalog Card No.* 74-77196

# Contents

# Map

# Preface

This book arises from a research project, begun in 1971, on the role of environmental pressure groups in the political system. We are indebted to the Nuffield Foundation, the Social Science Research Council and the University of Keele for their financial support of our research.

All but two of the chapters are being published for the first time· Chapter 3, 'The National Smoke Abatement Society and the Clean Air Act (1956)' by J. B. Sanderson, is reprinted with amended notes from *Political Studies*, vol. IX, 1961, by permission of the author and the Clarendon Press, Oxford. Chapter 6, 'The Minister's line: or, the M4 comes to Berkshire' by R. Gregory, is reprinted with amended notes by permission of the author from *Public Administration*, vol. 45, Summer and Autumn issues, 1967. Chapters 4 and 5 are based on M.A. and Ph.D. theses presented to Manchester University in 1966 and 1969 respectively. The remaining chapters have been written especially for this book.

Our thanks are due to John D. Lees, for reading and commenting upon chapter 9; to Stuart Brookes, who helped us at various stages in the preparation of the typescript; to our respective wives, Jennifer and Anne, for their help with the proof-reading; to Mr G. Barber, for drawing the map for chapter 4; and to Mrs A. Millward and Mrs A. W. Holmes, for undertaking the typing.

November 1972

# Chapter I
# Introduction

> You can't put a shovel to the ground anywhere on this planet any more without a Society Against surrounding you and saying: 'Fie!'
>
> John Crosby, *Observer*, 28 February 1971

The adoption of the somewhat incongruous term 'moral pollution' as the main target in the 1972 *Festival of Light* illustrates how readily environmental pollution has been accepted as a major problem. If we take the interest of the mass media as another index, concern for the environment would seem to have diffused throughout the population. Not only are environmental problems newsworthy to the extent of having feature articles in Sunday colour supplements and their own television programmes, but the environment has insinuated itself into such programmes as *The Archers* and *Softly Softly* and has become entertainment. Perhaps the ultimate act of trivialization is the manner in which advertisers have been quick to jump on the bandwagon. Some extol the high environmental pedigree of their product – like the insecticide for pollution-conscious gardeners advertised as 'the harmless killer' – while some are merely exploiting the current interest – like the vodka advert which managed to combine moral and environmental pollution in its 'clean up the countryside' advertisement.

It is not possible to chart the growth of media interest in the environment without a detailed content analysis, but in chapter 3 J. B. Sanderson gives some indication when he notes that the press have paid ever increasing attention to air pollution in post-war years and that the National Smoke Abatement Society, as it was then called, received only 60 newspaper cuttings a month in 1946. By 1951 the figure had risen to nearly 200, and by 1954 they were receiving 1,000 cuttings each month. Indeed, the volume of cuttings has since become too great to handle and the society has cancelled its order for press cuttings. Although press interest in air pollution declined after the passing of the 1956 Act, the International Press Cutting Bureau were still cutting 330 items per month by 1968 and this figure had doubled by 1971. In the first quarter of 1972 the bureau filed cuttings at the rate of over 1,400 a month on all aspects of pollution.

Now, all this attention to the environment, including the immense coverage given by the media, could be misleading. The possibility that the issue is almost exclusively of interest to what one might call the intelligentsia should not be overlooked. There are many examples of the communicators presenting as generally accepted what is only generally accepted by themselves – the wisdom of abolishing capital punishment being the most obvious. Indeed, Mr Anthony Crosland has suggested that much of the conservationist lobby is class biased and reflects a set of middle- and upper-class value judgments.[1] There is little doubt that the membership of the environmental lobby is mainly middle-class in character (any voluntary association tends to be dominated and run by the middle- or lower-middle classes[2]) but the assertion that its values are not shared by the population at large is more questionable, though evidence on either count is rather limited.

While the polling organizations have been busy producing major studies for private clients (presumably the big industrial concerns involved in mining and petro-chemicals, among others), relatively little has been published, and the academic world has not so far made up the deficiency. What evidence there is tends to refute Crosland's jibe about the middle class. Surveys conducted in January and April 1972 by National Opinion Polls showed that on each occasion about 2 to 3 per cent of the population regarded dealing with pollution as being the *single* most important problem then facing Great Britain. Of these, 46·9 per cent were manual workers in the first survey compared with 64·3 per cent in the second, suggesting that working-class involvement was, if anything, increasing. A note of caution should be sounded here: the numbers involved on each occasion were very small (forty-nine in January and forty-two in April). However, the point is that the evidence so far does not support the Crosland thesis and the chances that future surveys will produce evidence of very low working-class involvement seem small. This point may be reinforced further by pointing to instances of activity in working-class areas such as the Swansea housewives' blockade of the United Carbon Black factory in February 1971,[3] the unofficial and official union activity leading to the closure of the Avonmouth lead and zinc smelter in January 1972, and the activities of groups like Clean Air for Teesside. Although, as the N.O.P. figures show, environmental pollution is not yet widely regarded as the single most important problem we face, there is evidence of widespread concern about the problem. A survey conducted by the Opinion Research Centre for the *Sunday Times*, in June 1972, showed that 13 per cent of the respondents felt that 'stopping pollution' was one of the *two* major issues facing the Government. It is interesting to note that more people mentioned this as one of the issues than mentioned 'tackling the housing problem' and 'providing better education'.

**The environment as a concept**

While concern for the environment still seems to have some novelty value, it would be totally misleading to imagine that campaigns to conserve, preserve, or purify are in any way recent phenomena. It is not our purpose to give a history of such activity, though Dr Palliser sketches the historical background to preservation in chapter 2, and further discussion from an historical point of view may be found in books recommended in the Further Reading section.

What is new is, on the one hand, the global approach to the problem – both figuratively and literally – and, on the other hand, the popularization of the issues by the mass media. The emergence of the concept of 'system' in so many spheres of intellectual activity has encouraged a global approach to many problems. In particular the attention attracted by ecology has emphasized that what had hitherto been regarded as disparate causes, championed by many different pressure groups, were in fact interrelated. This, in part, explains why there have been so many different organizations campaigning for aspects of the environment with so few of them covering a broad front. Those that do, for example the Conservation Society or Friends of the Earth, are of recent origin.

The problem has become a global one in the literal sense too. International agencies, especially the United Nations, NATO, and the EEC, have become aware of our responsibility towards our environment, and the international conference at Stockholm in June 1972 has emphasized the extent to which nations acknowledge the common problem.

It is not easy to identify the component issues which are now subsumed under the general heading of 'The Environment', largely because many of the problems are interrelated such that mutually exclusive categories cannot be satisfactorily devised. However, some general comments can be made. Three issues are of long standing and date back to earlier centuries. These are: the concern to protect wild life and plants, the preservation of landscape and buildings, and concern at the growth of water and atmospheric pollution. The development of modern transport, especially since the Second World War, has introduced two additional foci for pressure – the motor vehicle and the aeroplane. Each stands accused of excessive noise, air pollution, damage to buildings, and taking up too much valuable land. Concurrently with the emergence of the global approaches to the problem, widespread attention has been focused on two other component issues. First, the growth in population raises problems of food and water conservation together with secondary problems such as the dangers from using persistent insecticides to protect food crops, and is itself an issue once again. Second, economic growth raises problems of resource conservation and of waste disposal, particularly the disposal of plastics, chemicals and atomic waste, and this too has become a much debated issue.

## Campaigning for the environment

The purpose of this book is not to survey the extent of the threats to the environment – there is already more than enough literature which attempts to do this – neither is our aim to evaluate the arguments for or against economic growth, the use of pesticides, or the preservation of the countryside.

The underlying assumption of the book is that the problems of protecting and improving the environment are essentially political ones, a fairly obvious point which few writers seem to have grasped. What this book seeks to do is to examine the political activity surrounding some of these environmental problems. It is not possible to illustrate political campaigns in all aspects of the environment, partly through lack of space and partly because in some spheres the major battles are still being fought (for example, the problem of mining in the National Parks). We shall consider seven instances of campaigning for the environment which represent a reasonable range of environmental campaigns. The particular case-studies have been chosen because they became important public issues in themselves, because they represent problems which many people are likely to encounter, or because they tell us something interesting about the politics of the environment.

In the first case-study, Dr Palliser considers the perpetual problem of the preservation of the history and character of our towns and cities. Chapter 3 looks at the work of a national environmental group, the National Smoke Abatement Society, and J. B. Sanderson describes its role in the passage of the 1956 Clean Air Act. In chapter 4, H. R. Burroughes unravels the highly complicated web of conflicting interests which attempted to resist the siting of electricity pylons across the South Downs. In the next chapter, Dr S. J. Dolbey examines the politics of Manchester's attempts to obtain more water from the Lake District in the early 1960s. Then Dr R. Gregory describes the protracted battle over the siting of the M4 motorway. In chapter 7 we examine, with S. K. Brookes, the controversy over the so-called 'heavy lorries', while in chapter 8 we describe the impressive campaign waged by the Wing Airport Resistance Association against the selection of an inland site for London's third airport. Finally, in the Conclusion, we offer some observations on the preceding case-studies and upon the environmental lobby in general.

## Notes

1 A. Crosland, *A Social Democratic Britain*, Fabian Tract 404, 1971, p. 5. It should also be noted that we have made no attempt to distinguish between terms such as 'conservationist lobby', 'amenity lobby' or 'environmental lobby'. This is not to deny the importance of rigorous definition, but merely to say that the distinctions indicated would not be appropriate in the context of this book. The terms are here used interchangeably to refer to the same phenomenon.

2 For interesting cross-national data on organizational membership see G. Almond and S. Verba, *The Civic Culture*, Little, Brown, 1965, ch. X.

3 The housewives have since regrouped to fight the company again (see *Guardian*, 11 November 1972; *Observer*, 11 February 1973; *Consernus*, February 1973).

# Chapter 2

# Preserving our heritage: the historic city of York[1]

D. M. Palliser

> Then glance below the line of Sussex downs
> To stucco terraces of seaside towns . . .
> Such Georgian relics should by now, I feel,
> Be all rebuilt in glass and polished steel.
>
> Sir John Betjeman, 'The Town Clerk's Views'

**The problem in perspective**

With the advent of new threats to our environment – such as the power lines, giant lorries and airports discussed in later chapters – we may easily forget how many others have a long history. The Corporation of London has been battling against pollution of the Thames since at least 1372, and in Wren's London coal-fire smoke was so common that Sir Christopher was driven to washing St Paul's free of soot even before it was completed. What is new is not the existence of water and smoke pollution but their vastly increased scale, and the same is true of the erosion of the fabrics of our historic towns and cities, with which this chapter is concerned.[2] Rebuilding of the urban fabric has occurred in all periods, and within limits it is an inevitable, indeed a healthy process for a living community. What has become a major threat, however, is the vastly accelerated pace of change since the Industrial Revolution. A growing public awareness of the scale of this threat has led at length to pressures for a more controlled rate of change, with old and new held in balance – to conservation, in a word, rather than preservation.

It was during Victoria's reign that many national and local societies first became active in preservation, as did a few enlightened local authorities. Chester, for example, acquired powers of preservation over its city walls by Act of Parliament (1884). The pace of destruction, however, revealed that private, and even local public, initiative was not enough. Pressure built up for controls by general statute, though the British distaste for central direction put the country well behind France and Germany. The first Ancient Monuments Protection Act, 1882, was passed only after a nine-year resistance in Parliament by the defenders of unfettered private property; and for this reason all the early statutes had to exclude inhabited houses (and churches) from their scope. Not until the Ancient Monuments Act, 1913, did the State acquire the right to impose preservation orders, and this was largely at the urgent insistence of Lord Curzon, who had barely prevented an American syndicate from demolishing Tattershall Castle and exporting its best furnishings. Thus by 1914, when many British cities had been largely rebuilt without adequate record of what was replaced, the law could still protect only individual and uninhabited properties, mostly rural. Furthermore, the redevelopment of cities, with deeper foundations, was beginning to destroy much of the 'buried archives' of older towns – the surviving foundations of earlier buildings underground – again with very little record being made.

The story of urban conservation in Britain has been one long race to catch up with destruction on an ever-enlarging scale. For if the position was perilous by 1914, it has become far worse during the last sixty years, despite well-intentioned statutes; it could be said with little exaggeration – at least until 1968 – that destruction rose in a geometrical ratio while conservation followed arithmetically. The threat, as in the Victorian era, has been greatest of all in the centres of historic cities, usually though not always the areas formerly enclosed by walls. These are the most valuable areas historically, architecturally and archaeologically, for they are usually the areas of longest continuous occupation, and the richest in diversity of age and style of buildings. But the reasons which made these areas attractive to the medieval merchant often remain valid for his successors, and it is the old walled centres which are most heavily under attack by office blocks, multiple stores and traffic schemes.

With all this erosion of the urban fabric, what has been the response of the public? National and local amenity societies (some of the local ones will be studied shortly) have campaigned persistently for the control of development, by buying and restoring properties themselves, by persuading others to do so, and by urging new legislation. Among the national societies which have been founded for this kind of purpose in the last eighty years have been the National Trust (1895), the Ancient Monuments Society (1924), the Georgian Group (1937), the Council for British Archaeology (CBA) (1944), the Civic Trust (1957) and the Victorian Society (1958). The Civic Trust has played a particularly influential role in towns as an independent charitable trust promoting urban improvement and offering advice and support to local civic and amenity societies.

As against this voluntary activity, legislation relevant to towns has continued to be tardy, partly because of vested interests in private property, and partly because archaeology was until recently rural in its bias. After the Ancient Monuments Act, 1913, which had little impact on towns, there was no significant advance until the Town and Country Planning Acts, 1932 and 1947. These established local planning authorities, which were empowered to make orders for the preservation of buildings 'of special architectural or historic interest'. Thus inhabited buildings could be protected for the first time, although the maximum penalty provided for flouting a preservation order was only £100.

Meanwhile the Town and Country Planning Act, 1944, created machinery for the listing of buildings of special architectural and historic interest by a team of trained investigators. Their work was to classify 'buildings of such importance that their destruction should in no case be allowed' (Grade I), buildings whose preservation was 'a matter of national interest' (Grade II), and all other buildings of special interest (Grade III). The lists have proved invaluable not only as planning aids, but also as a safety-net against demolition, for two months' notice is required for any intended demolition of, or major alterations to, any building on the Statutory List (Grades I and II). The Statutory List for England and Wales, slowed down by a staff reduction in the Labour Government's economy cuts of 1950,[3] was completed in 1968, when it included 116,000 buildings. In 1969 revised lists were begun, taking into account the new emphasis on townscape rather than isolated buildings. The unsatisfactory Grade III has been abolished, and about half the buildings on it are being upgraded to II to enjoy statutory protection.

The planning machinery of 1932-47 secured a great deal, but not enough to cope with the growing dangers. Preservation was still normally enforced only on outstanding individual buildings, and not on areas of high quality as a whole, so that as recently as ten years ago the Minister of Housing could permit the first large-scale replacements in Georgian Harley Street and Wimpole Street, merely on the ground that *individually* the old buildings were not outstanding. Further, the £100 fine did not deter wealthy redevelopers from demolishing without permission, and in 1966 the owner of the finest old house left in Solihull demolished it in defiance of a preservation order; he was fined the legal maximum but is said to have made £45,000 profit on the site. Such events spurred on demands for new and stronger legislation; but even more important as a turning-point was 'the sack of Worcester' in the early 1960s – the total rebuilding of a large area of the historic city centre, with the Corporation's blessing, and with very little attempt to preserve the old buildings or to record the archaeological evidence before its destruction. The outrage prompted a brave and angry journalist into an attack that gave considerable impetus to the growing movement for urban conservation.[4]

At just this time a government report by Mr (now Professor) Colin Buchanan and others, *Traffic in Towns* (1963), focused attention on another growing threat. In the course of a wide-ranging survey, they found that some smaller towns could accommodate the increasing volume of traffic for the foreseeable future (2010 was taken as a terminal date) with only relatively small changes in the road network, but that larger cities would need either redesigning on a vast scale, or severe restrictions on free traffic movements. Historic towns were studied as a special category, and a case-study of Norwich revealed that the 'environmental capacity' of much of the old city was already 'grossly exceeded'. The writers called for an end to the policy of catering for all expected traffic demands in such towns, and insisted on limiting the volume of traffic if the urban fabric were not to be destroyed.

The report became a catalyst for all those fears about historic towns which had been hitherto scattered, latent and relatively ineffective. In 1964 the CBA published a general statement about historic towns in the light of *Traffic in Towns*, followed the next year by a list of historic towns in Britain where planning and redevelopment needed special care. No less than 324 were listed, even though towns already damaged beyond recovery were omitted to avoid overloading the list. A second list selected fifty-one from the larger number as being objects of national concern. And in 1966 a third CBA report outlined the way in which planning should allow all aspects of a town's material history to be either conserved or at least recorded. Many of the fifty-one selected towns were already under threat of major redevelopments, and the Labour Government was moved to act under pressure from enlightened public opinion and from a few influential politicians. Notable among the latter were Lord Kennet, who in 1966 became a Parliamentary Secretary with special responsibilities for historic buildings under the Minister of Housing and Local Government (then Mr Richard Crossman), and Mr Duncan Sandys, a Conservative ex-Minister of Housing and the shrewd founder-chairman of the national Civic Trust. In 1966 the sympathetic Mr Crossman, in conjunction with the local authorities concerned, launched five pilot studies on historic towns selected from the short list of fifty-one, 'to discover how to reconcile our old towns with the twentieth century without actually knocking them down'. Shortly afterwards, Lord Kennet created a Preservation Policy Group to consider the pilot studies and historic towns in general, and to consider what legal, financial and administrative changes might be necessary. The group, under his chairmanship, comprised a panel of distinguished outsiders (the Winchester Town Clerk, the Norwich Planning Officer, an architectural historian, a professor of economics, and so on) as well as civil servants, and was to make recommendations to the Minister of Housing.

Meanwhile, Mr Sandys, putting his experience as a former minister to good use, promoted a Private Member's Bill to tighten up the preservation law and to extend its concept from individual structures to whole areas. With a praiseworthy bipartisan approach, Mr Crossman provided the help of the Ministry of Housing, and there resulted the Civic Amenities Act, 1967, which required local planning authorities to designate areas of special architectural or historic interest as Conservation Areas. Within such areas, all significant proposals for development had to be advertised in the local press and to be made available for public inspection. Next, under the impetus of Lord Kennet and the Preservation Policy Group, the Government itself strengthened conservation law in its Town and Country Planning Act, 1968. It encouraged participation in the local planning process by amenity societies and the general public, and blocked several loopholes in the existing law. Building Preservation Orders were abolished, and it was made an offence to demolish or damage any listed building without written consent – punishable by imprisonment of up to a year or by a fine proportionable to the offence.[5]

## York as a case-study

The pilot studies chosen by Mr Crossman in 1966 were of Bath, Chester, York, King's Lynn and Chichester, of which the first three are all major pre-industrial cities with a complex urban fabric. This applies especially to York, one of the half-dozen most important provincial cities from Roman times to the eighteenth century, and the richest of them to survive, with the possible exception of Norwich. Despite much erosion in the past two centuries, it still boasts its Minster (the largest medieval cathedral in Britain), nineteen medieval or partly medieval churches, almost its complete circuit of medieval stone walls, and nearly 1,000 listed buildings. The survival of such wealth is owing chiefly to the fact that industrialization came late to the city and on a modest scale, so that the nineteenth century caused relatively little damage, and also to York's good fortune in escaping the fate of Coventry and Exeter during the bombing raids of the last war. Yet even at York the cumulative toll of 'progress' has been heavy over the past century. Too many fine houses have made way for offices and stores out of scale with their surroundings, often poorly designed into the bargain. Too many frontages and corners have been set back to ease traffic congestion, and wide new streets have cut ruthlessly across the old pattern, only to find that, by an extension of Parkinson's Law, traffic increases to fill the roadspace available. Too many landmarks of character have gone, among them four redundant parish churches;[6] and the west bank of the Ouse has come to resemble part of the City of London, with the delicate spired church of All Saints dwarfed by a huge hotel and insurance offices.

Many of these difficulties are faced by all English historic cities, but two aspects of the Corporation's situation can be suggested which may have made the problems at York particularly acute. One is that the city, like Venice, is trapped by the legacy of its own past; it is a treasure of national importance, yet the Corporation's income is relatively modest, and it does not believe it practical (or, for that matter, fair) for the citizens to pay the entire cost of conservation without some help from the central government. Another may be the even balance of local politics; for although, until 1964, the city has normally elected a Conservative MP, the distribution of power on the City Council has usually been much more even, and it might not be unreasonable to deduce from the Council's activities before the late 1960s that either party, when in power, tended to put economy with the rates, and the short-term interests of motorists and shopkeepers, a long way ahead of aesthetic and cultural values, and even of their long-term self-interest of promoting tourism. If, as will be suggested, there has recently been a marked change of direction, that is owing in large part to a recent change of climate in the city.

Some feeling for conservation can be traced back in York for at least four centuries.[7] In 1596, for instance, the Corporation successfully pleaded with the Crown to preserve Clifford's Tower, which they called 'the most especial ornament for show and beautifying of this City . . . York Minster only excepted'. Gradually there grew up a broader appreciation of the whole urban fabric, notably by the Yorkshire Architectural and York Archaeological Society (YAYAS), founded in 1842, and the main force for conservation in the city until 1946. Among many other activities, they called what was believed to be the first public protest meeting ever held in York on a preservation issue,[8] and in 1927 they won a legal action against the Corporation, which was trying to destroy the Cholera Burial Ground for a traffic scheme. The Corporation itself could be conservationist: in 1934 it began to convert a disused Georgian prison into a home for the remarkable collection of shopfronts, workshops and other items by Dr J. L. Kirk, and so began what has grown into a famous folk museum drawing over half a million visitors every year. In general, however, the Corporation did not (indeed could not) take an active part in conservation, and it was therefore fortunate that one of its most active members was Alderman J. B. Morrell (1873-1963), who was instrumental not only in creating the Kirk Museum but also in changing the climate of civic opinion. In *The City of Our Dreams* (1940) he sketched what would now be called a conservationist plan for York's centre, pleading among other things for traffic restraint in the centre, coupled with an inner ring road encircling the walls.

Meanwhile two more voluntary societies strengthened conservationist pressure in York. There had been a growing appreciation of the medieval, Tudor and Stuart legacy for a century and more; but the Georgian style – one of the glories of York – had been so far unfashionable. A change of heart was signalled by the founding in 1939 of the York Georgian Society, which has since worked hard to stimulate an interest in art and architecture of the period 1660-1820, and to preserve threatened buildings of that age. Even more important has been the York Civic Trust (not to be confused with the national Civic Trust), founded in 1946 with the triple aim of preserving the past, raising environmental standards and encouraging good design. It was lucky in having influential patronage from the start: Alderman Morrell was for many years its chairman, and Dean Milner-White (1884-1963) played a prominent part.

In 1948, the year the Corporation received its new powers as a local planning authority, there was published a report it had commissioned on the walled central area, by a team led by Professor Adshead.[9] This recommended, among other policies, restoring dilapidated areas like the Shambles and replacing slums by new flats. Most notable were their traffic proposals. Within the walls, two new streets were to be cut through for traffic, but three existing streets were to be converted for pedestrian use only. The report also endorsed the Morrell inner ring road, circling the walls at a distance of 250 yards, and it further recommended an outer ring road beyond the built-up area. During the next twenty years, the City Council put some of the more limited Adshead proposals into effect, reserved some for future action and ignored others. For instance, it consistently postponed closing Deangate, which would be unpopular with motorists, in spite of the possible danger to the Minster fabric. And the recommendation for pedestrian streets waited until the closure of Stonegate in 1971, by which time Norwich had overtaken York as the pioneer in this approach.

During these twenty years, the pace of conservation was set by the York Civic Trust rather than the Corporation. It did not so much supersede the older amenity societies as co-operate with them, and the three main ones (the Yorkshire Philosophical, the Georgian and the YAYAS) were all represented on the Council of the Trust, which became to some extent an umbrella organization. Until the mid-1960s it was a relatively small society: it had only 200 members in 1961 (as against over 600 in the Georgian), but it was acquiring and spending funds on a large scale: with a membership subscription of only a guinea, it yet enjoyed an income of over £1,000 at that time. It achieved this in part by securing substantial gifts and grants from individuals and trusts: for instance, it was able in 1957 to start a Fund for the Preservation of the Mediaeval Churches of York as the result of a large anonymous donation. The Trust was also contributing to the cost of Corporation schemes, and carrying out others itself, such as the restoration of a fourteenth-century house, which was so arranged that a substantial sum is now received from lettings, the balance, after repairs have been met, going to swell the Trust's general funds.[10] Most significantly, the Trust as early as 1949 sponsored an academic committee, in anticipation of the establishment of York University thirteen years later. The committee were able to restore a medieval guildhall as a record office and a redundant church as an Institute of Advanced Architectural Studies, both of them projects since taken over by the university.

In the last decade, fortunately, the Corporation has itself been taking a much more positive approach to urban conservation. In 1964 a member of the City Engineer's staff produced a pioneer booklet on the preservation of historic buildings, which for the first time advocated the conservation area principle at York.[11] After discussions with the Ministry of Housing and Local Government, the Corporation in 1966 launched a Historic Buildings Grant Scheme which was, and remains, the largest in the country. Protection is extended to groups of buildings and to streetscape, as well as to individual properties. Owners unable to finance repairs and restoration can apply for a 50 per cent grant which is provided jointly by the Corporation and the central government, and in the first six years of the scheme over £43,000 was paid out in this way. The success of the project allowed the Corporation to respond quickly to the Government's call for Conservation Areas in 1967, and four such areas were designated promptly in 1968, the largest of them comprising the whole historic core of York.

## Esher and after

These activities by the Corporation and Civic Trust seem to have been instrumental in Mr Crossman's decision, in 1966, to make York the object of one of the five pilot studies, for evidence of self-help by the cities concerned was one of the factors governing his choice. A team of consultants under Lord Esher was commissioned to study York, and early in 1969 they produced a lengthy report[12] to the Minister and the Corporation, on the city within the walls (the rest having been excluded from their brief). They defined five objectives at the outset:

1 That the commercial heart of York should remain alive and able to compete on level terms with its neighbour cities, new or old.
2 That the environment should be so improved by the elimination of decay, congestion and noise that the centre will become highly attractive as a place to live in for families, for students and single persons, and for the retired.
3 That land uses which conflict with these purposes should be progressively removed from the walled city.
4 That the historic character of York should be so enhanced and the best of its buildings of all ages so secured that they become economically self-conserving.
5 That within the walled city the erection of new buildings of anything but the highest architectural standard should cease.

The Esher report, though drawing rather more on the 1948 Adshead report than it acknowledged, was in the main a bold and novel attempt to tackle York's major problems as a whole, and especially the related problems of blight and of traffic. Accepting that the walled city was too large (263 acres) to ban traffic altogether, the Esher team defined an 'inner enclave' of about half that size, in which the best townscape was concentrated. Within this enclave servicing vehicles were to be strictly controlled, and some streets to be closed to traffic entirely, while the private cars of non-residents were to be completely barred from the walled area, except for prescribed routes leading to new multi-storey parks. These traffic proposals were, of course, only a means to an end, and the full Esher recommendations cannot be summarized briefly without distortion. Suffice it to say that, in the view of many architects and planners, the report attained all of its objectives. In particular, it showed that by 'infilling' decayed areas with compact new housing, and by renovating the empty storeys of older structures, the city within the walls could comfortably house 6,000 people instead of the present 3,500. The renovation of old buildings was found, contrary to popular belief, to be not necessarily costlier than new building.[13]

Cost, of course, is the crux of such a plan, and Lord Esher commissioned Professor Lichfield of University College, London, a planning consultant, to estimate the costs to the Corporation of carrying out the entire report. These amounted to a £1·1m. net loss incurred in new building, after selling off the freeholds, and a further £1m. spent on conversion, repair, restoration and environmental improvements. The financial proposals were summarized as follows:

> First, on the net cost side, we have a grand total in capital terms of £2·1m. Assuming . . . that redevelopment in the interests of conservation will rank for grant on the same basis as redevelopment in the interests of modernisation, we have anticipated a 50 per cent central government grant for redevelopment and a 40 per cent local authority grant for repairs and conversions. We have then converted capital cost to annual cost and arrive at an annual cost to the local authority of £90,000, which in York is a 6d. rate. If to this we add an allowance of £10,000 p.a. for running a Conservation Section . . . we have a total annual cost of £100,000.

The total capital cost of £2·1m. can be placed in perspective against the £2m. for which the Minster asked in 1967 towards restoration, and which was raised in full within five years by public appeal.

Publication of the report produced a sharp initial surge in enthusiasm among city councillors and the general public, in spite of its very high price of £7. The local *Evening Press*, by producing a short version at 7s. 6d., helped considerably in taking it to a wider public. The local amenity societies all strongly endorsed Esher's main proposals. But despite all this, and despite Esher's insistence that 'it is an "all or nothing" situation, in which half measures would be a total waste',[14] the councillors seemed inclined to plump precisely for half measures, as they had earlier done with the Adshead report. In October 1969 the Corporation published its officers' comments on Esher, in a report described by the *Architectural Review* as 'ill informed, misleading and tendentious'. 'It is not encouraging', chided the *Review*, that York, 'likely to provide the test case of whether local authorities as well as the Government can be persuaded to take urban conservation seriously', had not backed the report 'in the way it deserves'.[15]

There can be legitimate disagreement about the detailed Esher proposals, and no corporation need feel obliged to endorse a commissioned report *in toto*; but fundamentally the Council's negative attitude sprang from its unwillingness to give conservation a sufficiently high priority, in spite of rebellions by individual councillors. In 1971-2, however, a more hopeful situation came about, when a reshuffle within the ruling Conservative party made a conservationist, sympathetic to Esher, leader of the Council. This year at last saw the implementation of some of Esher's more modest proposals. Exhibition Square was paved over, and Stonegate barred to traffic for most of the day. More importantly, the city at last acquired a separate planning officer,[16] and Lord Esher was commissioned to carry out the Aldwark scheme, one of his own proposals for renovating and 'infilling' a decayed central area. During the same year one other project was initiated which went beyond the Esher proposals. The pace of development, and especially the proposed inner ring road discussed below, was alarming the archaeologists. The CBA, in conjunction with the Yorkshire Philosophical Society, commissioned a survey on *The Archaeological Implications of Proposed Development in York*, which found that threats were posed by development in the decade 1972-81 on a very large scale, especially by the inner ring road and by the Aldwark and other redevelopment schemes, however desirable they might be in other ways. The report led, early in 1972, to the creation of a York Archaeological Trust, which intended a ten-year programme of systematic excavations, financed by public and private sources, on the model of the successful similar programme at Winchester. Financial contributions were promised, initially for the first year, by York Corporation and by the Department of the Environment.

For this changing climate of opinion in the city, how far can the amenity societies take the credit? They are certainly not large either in numbers or financial resources, whatever their indirect influence. The Philosophical Society has oscillated since the war between 100 and 500 members; the Georgian Society has some 500 members, as against over 600 ten years ago; and only the Civic Trust has shown sustained growth in the past decade, from 200 in 1961 to 750 in 1971. There is considerable overlap of membership between these three and the YAYAS, and membership of all seems predominantly middle class. Active leadership has come since the war mainly from a small group of architects, teachers, churchmen, city councillors and other professional people – the Chairman of the Civic Trust, for instance, Mr John Shannon, is a solicitor, JP and former councillor, while the President is the Archbishop of York. The financial resources of the societies are modest. The Georgian Society, for instance, with a guinea subscription, has a turnover of less than £1,000 p.a., and the YAYAS of under £400. The largest spender on conservation is the Civic Trust. The annual subscription for ordinary members was a guinea until 1970, when it was raised to £2; but the Trust also benefits from generous donations from individuals and corporate bodies, and has received grants from the Pilgrim Trust. Consequently, in its first twenty-five years (1946-71) it was able to spend £22,000 on grants in aid, including nearly £7,000 on church repairs and the Minster Fund, £5,000 towards Lord Esher's fee, and a multitude of smaller payments for building restorations, public paving, commemorative plaques, and purchases of furnishings for the Mansion House and pictures for the Art Gallery. In some cases it has been able to prime the pump by laying out the capital cost of a project (e.g. floodlighting and fountains) and persuading the Corporation to undertake the running costs; in others it has raised half the cost of a building's conservation and persuaded the owners (sometimes the Corporation) to pay the rest.

It is clear that the York amenity societies have achieved successes disproportionate to their size and resources (though not nearly as many as they would wish). This is partly because they tend to adopt a similar stance on major issues, through a shared active membership and through representation of the older societies on the Civic Trust. The Trust itself has also achieved success by ensuring good publicity in the local press and elsewhere, and by creating a generally good working relationship with the Corporation, in which the chairman's experience as a former councillor has been a considerable asset. The Trust does not admire the bitter public clashes which occur between societies and local authorities in some other towns, and prefers when possible an approach of compromise and of discussion in private. Sometimes, its critics feel, its laudable intentions work against each other, and in the past councillors have on occasion first learned of some '50 per cent' offers from the press; this has been resented as seeming to present a *fait accompli*. But in general the Trust has acquired a semi-official position which ensures that it is frequently consulted, while allowing it freedom of action. Perhaps it is for these reasons that the Corporation has not yet created any formal machinery for direct public participation in planning, as the Department of the Environment advises local planning authorities to do. It has, however, set up a Conservation Advisory Panel, which makes recommendations to the Planning Committee. Membership is by invitation from the Corporation, and the present chairman of the panel is Mr Shannon.

Attention has so far been focused on the city within the walls, and on the relative harmony between the various voluntary societies; but this is to neglect the contentious issue of traffic outside the walls and its relationship to plans for the centre. Lord Esher's brief was confined to the walled city, where he advocated severe traffic restrictions; but this of course had considerable implications for the rest of the city, since traffic barred from one area must either cease to move or find alternative routes. The Corporation therefore decided on an inner ring road on Morrell-Adshead lines, hugging the walls, a decision endorsed by the Ministry of Transport. This route would have done considerable damage to the urban fabric, especially on the west side of the city, so after further consideration the proposed road was on the western side moved further out. The Civic Trust and its affiliated societies are reluctantly prepared to concede the need for some inner ring road (and the outer ring road also projected), while being still not altogether happy about the revised route. It is not clear at the time of writing what will be the precise position of the various societies during the Public Inquiry on the ring road in October 1972. They agree with the Corporation that there must be an inner ring somewhere, and that the revised route is a great improvement. Their view was given weighty support in October 1971 by Lord Esher, who, since his report had excluded the extramural area, made his views known in a letter to the Civic Trust:[17]

> The traffic which must be cleared from the walled city is in great part . . . cross-town rather than through traffic. An outer by-pass will deal only with the latter minor component, and . . . some form of inner loop or ring road is therefore necessary. . . . I do not believe that it would be safe to wait for new forms of public or private transport to solve the problems of transportation in the walled city . . . [and] the best thing the city can do is to get on with both the outer and inner rings (as now amended) as rapidly as possible.

Nevertheless, the societies' consensus with the Corporation over the roads is an uneasy one. In February 1972 society meetings were held specifically to debate the inner ring: the Georgian Society passed one resolution supporting the road in principle, and another condemning the Bootham Park section, while the Civic Trust more forcibly voted that it could not support the proposed road unless the part proposed to follow the walls was re-examined. The debates revealed considerable division: some members were very unhappy about the road, but Dr Nuttgens, a member of both societies and a persistent and eminent architectural critic of the Corporation in the past, gave the road general support. Yet other members of the Georgian Society expressed the more traditional view that their role was to conserve and study Georgian buildings, and not to pronounce on road schemes.

Meanwhile, however, the general consensus between the Corporation and the amenity societies has provoked the creation of other groups of a rather different kind. The first of these was the York Group for the Promotion of Planning (YGPP), created in 1965 by Dr Nuttgens and others, as an informal pressure-group for comprehensive planning of high quality. Although it soon became a registered charity and amenity group, it has remained a small body (30–40), and has tried to fulfil a different role from the Civic Trust both by including in its study groups the areas outside the walls, and in being more outspoken in criticism of the Corporation: it was they, for instance, who first suggested that the city needed a separate planning officer, and that greater public consultation on planning was necessary. At the end of 1971 some members of the YGPP, worried about the inner ring road, decided to set up a new transport pressure-group with mass membership, 'York 2000'. With a city councillor as chairman and a university professor's wife as publicity officer, it has set itself to press for the examination of alternative transport proposals, better public transport, and so on, since it is not convinced that an inner ring is necessary. It has cast aside the muted tone of public criticism of the Corporation adopted by the older societies, and has been much more flamboyant and imaginative in its publicity, setting out to create mass membership through a low entrance fee (10p). During the first half of 1972 it sponsored processions through the city centre, took a full-page advertisement in the local newspaper, and raised money by jumble sales, auctions, parties and dances. It quickly attracted a wide cross-section of residents and visitors, skilfully exploiting the membership drive for better publicity (Sir Mortimer Wheeler became the 2,000th member and Mr J. B. Priestley the 3,000th). Membership in September 1972 stands at 8,600, and the group has raised over £6,000 towards the £7,500 needed for representation at the ring road inquiry. Professor Lichfield, invaluable for his experience with the Esher team, has been retained by them as a consultant.

At present, therefore, the future of conservation in York remains in the balance. There are at last hopeful signs of a more positive approach within the walls, but what is to be the price of it? Much archaeological evidence will inevitably be destroyed by otherwise desirable developments, and it is to be hoped that public finance will continue to be forthcoming for the necessary excavations. Traffic restraint within the walls will, in the Corporation's view, inevitably involve not only a partial outer by-pass but also an inner ring road, which is bound to be in some sense destructive whatever revised route may be adopted. One has only to see the inner ring at Chester to appreciate how disastrous such a road can be if not routed and designed with very great care.

## A national policy?

No one historic city can be considered 'typical', least of all York with its unusually rich and complex urban fabric. Its complexity has meant that the city now faces almost the whole range of possible threats to historic cities in this final third of the twentieth century; but no town study has meaning in complete isolation, especially as national control of planning and purse-strings becomes tighter. Indeed, in 1974 the county borough status enjoyed by cities like York is due to disappear, and with it the last attempt of communities 100,000 strong to run their own affairs. It is therefore necessary to revert to the national situation as a whole, for it is only by nationwide action and by nationwide financing that our historic cities can be saved.

To some extent the national outlook is rather more hopeful than it has been for a long time. The 1967 and 1968 Acts introduced the concept of Conservation Areas as urban and rural districts worthy of conservation *as a whole*, and by mid-1971 1,500 such areas had been designated by local planning authorities, over half of them in towns. As we have seen, the York Corporation designated the entire intramural city as one Conservation Area. All of this has had a considerable effect in stimulating a greater interest in the environment of these areas, and even in some cases a purely material return as property values have been enhanced. But the great omission of the current legislation is that it lacks teeth; an authority is not compelled to designate any such areas, and even when it does so, this means only that planning proposals within those areas must be advertised and open to public comment. This is not a plea for a rigid control on building demolitions; even a Grade I listed building in a Conservation Area may sometimes need to be demolished, for conservation involves a healthy balance of change and continuity, rather than ossification. But some stronger statutory protection for the Conservation Areas and for outstanding individual buildings is surely overdue, since the dice are loaded so much in favour of change in valuable central districts that there seems little immediate danger of going to the other extreme. It is still possible for private redevelopers, without breaking the law, to have their way in the end: if a whole building is ordered to be preserved, then the structure can be quietly but deliberately allowed to decay; if only the façade is protected, then the core can be so rebuilt as to make the front dangerous.

Nevertheless, the problem is one of finance as much as of statutory protection. The feeling dies hard in many quarters that any restriction on an owner's right to adapt or demolish a building is at best a necessary evil, for which monetary compensation should be payable. In very many cases, however, planning restrictions on development are affecting not house-owners of moderate means, but businesses wishing to redevelop city-centre properties on a large scale. If such developers acquire areas which they are then refused permission to redevelop, especially if the buildings concerned are listed, have they in fact any legitimate grievance? More hardship is caused where an owner of moderate means possesses a single historic property which he cannot freely adapt for his business. There is a system of public grants available for maintenance of historic buildings, but it is generally cumbersome and inadequate in scale, though there are exceptions like the very successful scheme at York. What is needed is a large-scale flexible scheme under which finance will be available in historic cities for all kinds of help – for conserving fabrics, for redecoration sympathetic to the surroundings, for conversion of decaying upper floors into flats, and so on. In addition, owners of listed buildings could be granted tax relief on the repairs they make.

Some local authorities, considering their limited financial resources, have achieved useful results. Chester, alone among English towns, raises 1p rate for conservation; as a result, it can offer owners of historic buildings a grant covering half the cost of restoration. York, jointly with the Government, similarly offers 50 per cent grants. But public expenditure of this kind, however laudable, does not meet the whole need. Large-scale schemes are needed which will raise money as well as spend it, so that town-centre conservation can become largely self-financing. Professor Lichfield's proposals for York represent one possible approach, reducing a total cost of £2·1m. to £1·1m. This scheme implied local initiative, though with a 50 per cent Exchequer grant; but Donald Insall's 1968 report on Chester suggests a bolder approach: the creation of a National Corporation for Historic Towns, somewhat like the New Towns Commission. This proposal, despite growing support in some quarters, has not yet been acted upon, for it has aroused bitter opposition from many of the local authorities concerned. They argue that such drastic powers are undemocratic, since they take power from the elected local representatives, and are also unnecessary, since the present elected authorities can manage quite happily given increased public support and financial help from the Exchequer. The case of Winchester has been especially stressed, as one where a small non-county borough has co-operated in a ten-year programme of fruitful archaeological research (1961-71).

But one swallow does not make a summer. *The Erosion of History*, a study on urban redevelopment published by the CBA in May 1972, shows that out of 702 historic towns in England nearly a fifth (127) are so seriously threatened that most of their archaeological evidence will be destroyed within twenty years, and many others face destruction on a smaller scale. Altogether 457 towns are threatened by serious redevelopment, but there is at present adequate investigation of below-ground archaeology in only seventeen, and adequate study of surviving buildings in only nine. Only one urban authority (Winchester) is spending as much as a 0·2p rate on excavations.

With such a situation in the towns (to say nothing of the countryside) it is not surprising that 1971 saw the creation of 'Rescue – a Trust for British Archaeology', to raise money for surveys and rescue excavations, and to campaign for tougher legislation. Its dramatic advertisements proclaimed that 'as the golden remains of Tutankhamun are exhibited in magnificent splendour, Britain's priceless past is being bulldozed away'. Partly in response to this pressure, the Government agreed in January 1972 to increase its annual grant for rescue excavations by almost half.

Rescue's aims can be met in part by limited legislation, compelling developers, for instance, to grant access by archaeologists to cleared sites, and where necessary enforcing the postponement of redevelopment. Indeed, after a parliamentary outcry in the spring of 1972 over the Baynards Castle site in London, the Government announced that legislation for powers of compulsory excavation was being considered. But this is only one aspect of the threats to historic towns, and the root of the problem remains how to find the expertise and finance to plan historic town-centre conservation on an adequate scale. The present local planning authorities have largely failed in this task. It is true that in 1974 a massive reorganization will make local authorities larger and give them greater resources, but the Government's present intention is to give wide planning powers to the second-tier authorities rather than confine them to the larger county authorities, and so decisions of great importance to historic towns will continue to be made by relatively small authorities of often limited vision. Is there any alternative, if town-centre conservation is to be adequate, to a national authority with strong powers and adequate State finance, specially equipped to deal with historic towns? An increasing body of public opinion would argue that there is not, and that the fifty-one historic cities listed by the CBA are national as well as local responsibilities, just as Venice is coming to be seen by the Italian Government as a national responsibility. The financial burden, according to some advocates of this policy, would not be too heavy when shared among all taxpayers. Lord Esher has argued that the national stock of buildings worth conserving is equivalent to only four months' new building at the present rate. He maintains, further, that the majority of the fifty-one towns can become self-supporting in terms of the finance of conservation; those towns needing special government help could be few, and 'the whole could be completely restored and secured for less than the cost of a few miles of urban motorway'.[18]

Many local councillors would argue that this solution is a negation of local democracy. But there are two reasons why the present structure of local government, admirably democratic in theory, fails to reflect adequately the growing public concern over the environment in general and over conservation in particular. One is the interdependence of local areas: visitors to a city, who often contribute greatly to its wealth as shoppers, consumers and tourists, have a legitimate right to a say in its central conservation policy, although they have no votes there. (The logic of the total autonomy plea by some councillors would dictate equally their right to pollute surrounding areas with smoke or chemicals, provided that their own area was unaffected.) Another reason is that local politics, like their national counterpart, still tend to crystallize round issues unconnected with the environment, so that local elections do not usually reflect these public concerns at all adequately.

We thus come round to our starting-point again: public concern. To some extent the national and local amenity societies, which have played such a valuable role, will continue as important channels of public participation. The national and local Civic Trusts, and bodies like the National Trust and the CBA, have won the right to consultation by the State and the local authorities on many issues. Some societies can do valuable work however little legislation there is; the National Trust goes on buying more properties to save them for public enjoyment, and will no doubt continue to do so whatever the scope of future laws on conservation. But these societies may have retarded legislative changes by their very successes; there is always a temptation for some politicians to make private initiatives an excuse for averting government action, as though such initiatives can ever be adequate on their own. And even though a growing number of local authorities are following government advice to build public participation into their planning structure, this often takes the form (as at York) of a committee representing local societies rather than the public at large.

Local authorities' defence of this approach is usually that they are, by right of election, representative, and that there is no need for a 'direct democracy' approach to public participation. But in the real world of local politics this plea seems somewhat hollow, especially when the public concerned are remote from their elected representatives. York, with 105,000 people, is relatively small and accessible to public opinion; but what of the difficulties of public feedback in, say, central London? A fierce battle in 1971-2 over the redevelopment of Covent Garden has come about largely because the planners are not the local borough but the Greater London Council, which seems very remote to many local residents. Its views seem to have been adopted after much thought on urban conservation and sympathetic redevelopment, but the information gap between residents and planners could scarcely be wider. Whatever forms of public participation are adopted by the new provincial local authorities after 1974, they must surely be designed to avoid this kind of situation.

If public participation is essential for effective urban conservation, so too is resolute government action, central as well as local. Lord Kennet has recently given an engaging account of the Labour Government's record on conservation between 1966 and 1970, and of his own part in it.[19] He concludes with the Government's announcement, just before the 1970 election, that they accepted the main recommendations of the Preservation Policy Group, which reported to the Minister that pilot projects should be carried out in the four selected cities[20] (like the Aldwark scheme at York), and that the government should meet 50 per cent of the cost of general conservation schemes in historic towns. The record of the Conservative administration since 1970 has been uneven from this point of view. The creation of a Department of the Environment under Mr Peter Walker has allowed a rather broader approach to urban planning than was possible before. On the other hand, when Mr Walker finally passed judgment on the report of the Preservation Policy Group in January 1971, he announced that only some of their recommendations would be accepted. A decision on the all-important plan for 50 per cent Exchequer grants was declared to be 'premature', and apparently it still is. Lord Kennet has called this decision, with pardonable exaggeration, 'the first setback in the development of public policy on preservation since it began in 1873'.[21]

Certainly time is very short if the four selected cities – and the many other historic towns – are to be saved. This was demonstrated by a recent investigation of Bath, where since the 1968 report on the city, a great deal of minor Georgian housing – the necessary foil to the larger set-pieces – has been demolished with the approval of the Corporation. Somewhat disconcertingly, the Town Clerk asked 'Is there some mystique that because houses are pre-1820 they ought to be preserved?' and declared that 'Our own young people demand the sort of city for living in that the kids in Birmingham, Manchester and Liverpool have'.[22] If all this can take place in one of the five 'pilot' cities – chosen partly because they were towns already showing some initiative on conservation – what may not happen in the others? A major battle lies ahead for conservationists if the Government is to be persuaded to save most of the fifty-one historic towns in time. Any forecast, however, could be in danger of erring on the side of pessimism; for if the threats to the towns are multiplying rapidly, so are the numbers who care about them and who will press for more action. A hopeful augury for the future, and a result of such pressures, was the announcement in July 1972 that the Government was increasing the budget of the Historic Buildings Council by 50 per cent. If Lord Kennet is right that the country's 'low-water mark in visual consciousness' was reached in the second quarter of the century, and that since 1945 the country has been 'steadily and rapidly emerging from a state of blindness',[23] then perhaps there is just time to save the cities.

**Postscript** (March 1974)

This chapter was completed during the summer of 1972. Since then there has certainly been a growth in support for conservation in many towns, with the founding of more rescue archaeology units and of many new amenity groups. At York, however, the situation remains almost unchanged. The outer ring-road has been started, but though the inner ring-road Inquiry was held in 1972, the Secretary of State for the Environment has yet to announce a decision on it. There is still only one 'footstreet', and the Aldwark area has continued to decay, though a government grant of £480,000 announced in March 1974 may at last resolve the deadlock between central and local government over implementing the Esher scheme. Nevertheless, despite some unfortunate demolitions and redevelopments, York continues to experience a growth in conservationist feeling, and the centre is in general receiving sympathetic care and restoration.

## Notes

1 I should like to thank those officers of the York Corporation and the York amenity societies who have taken the time to answer questions and to explain their policies; Mr John Harvey for much advice and information; and the editors. Any errors remaining in the text are, of course, my own.

2 The Germans had a word for it by 1904, *das Stadtbild*, recently copied by our own 'townscape'. I have preferred 'urban fabric' since it includes the archaeological remains below ground as well as the surviving buildings, street patterns, elevations and skyline. For the history of urban conservation, see John Harvey, *Conservation of Buildings*, John Baker, 1972.

3 See the *Architectural Review*, November 1970, p. 310, for an account by two of the scholars involved. Lord Kennet, however, tries to pin the blame on the incoming Conservative Government of 1951: Wayland Kennet, *Preservation*, Temple Smith, 1972, p. 50. His partisan approach is the only defect in an otherwise admirable account; for in truth neither party has a particularly good record on conservation. Lord Kennet himself was an untypical, though fortunately influential, Labour minister, just as Mr Sandys is an untypical Conservative MP.

4 Geoffrey Moorhouse, in the *Guardian*, 28 November 1964. Sir Nikolaus Pevsner's guide to *Worcestershire* in The Buildings of England series, Penguin, 1968, says of the central redevelopment: 'It is not easy to be fair to it.'

5 This and earlier Acts have now been consolidated as the Town and Country Planning Act, 1971. It should perhaps be added that the turning-point in the movement towards public participation was reached with the Skeffington Report, *People and Planning*, HMSO, 1969.

6 Space forbids adequate consideration of multiple church redundancies, a special problem of cities like York. Norwich, with thirty-two central churches, has set a good example by finding other uses for five so far. At York, a Commission on Redundant Churches, set up in 1964, has issued reports recommending uses for most of the surviving churches, and a York Redundant Churches Uses Commission, chaired by Mr Shannon, is now working on the problem.

7 Harvey, *Conservation of Buildings*, pp. 166-7, describes some Tudor conservation measures.

8 Over the siting of the City War Memorial (1922).

9 S. D. Adshead, C. J. Minter and C. W. C. Needham, *York: a Plan for Progress and Preservation*, York Corporation, 1948.

10 In 1971-2 the property (Bowes Morrell House, 111 Walmgate) brought in rent of £1,000. £208 was spent on various items, including a payment to its Repairs Fund, and the remaining £792 was transferred to the Trust's general account.

11 J. M. Hargreaves, *Historic Buildings – Problems of their Preservation*, York Civic Trust, 1964.

12 Viscount Esher, *York: a Study in Conservation*, HMSO, 1969, but dated 1968.

13 The report instanced a timbered house in Micklegate, converted and modernized for £3,669 to accommodate eight students. This worked out at £460 per head, against a University Grants Committee average for new building (in 1968) of £1,440.

14 *Listener*, 17 April 1969.

15 *Architectural Review*, December 1969.

16 Previously the City Engineer was also the Planning Officer.

17 Printed in *York Civic Trust Annual Report* 1970-1971, pp. 46-47. Lord Esher has since asserted that his report was 'highly critical of the ring road concept' and suggested an alternative (*The Times*, 15 September 1972). This seems rather too forceful a description of some very tentative criticisms in his report (pp. 52, 53). In any case, whatever his views in 1969, the letter quoted here is quite unambiguous.

18 *Listener*, 17 April 1969.

19 Kennet, op. cit., pp. 49-107.

20 Of the five pilot studies listed earlier, that for King's Lynn was abandoned at an early stage. Reports for Bath, Chester and Chichester, as well as York, were published in 1968-9.

21 Kennet, op. cit., p. 104.

22 *The Times*, 22 April 1972. A lengthy correspondence followed, in which the Corporation defended itself against a very weighty and vocal opposition.

23 Kennet, op. cit., p. 51.

## Chapter 3

# The National Smoke Abatement Society and the Clean Air Act (1956)

J. B. Sanderson

> Ships, towers, domes, theatres, and temples lie
> Open unto the fields, and to the sky;
> All bright and glittering in the smokeless air.
>
> Wordsworth, from 'Composed upon Westminster Bridge', printed in each issue of *Smokeless Air*, the journal of the National Smoke Abatement Society

Smoke abatement is within that class of subjects sometimes designated as 'non-controversial', not because no controversies are involved, but because political parties have no definite policies about it. Such a situation naturally leaves more room for pressure-group operations than one in which there are clear party lines. The purpose of the present article is to give some account of the pressure-group activity surrounding the Clean Air Act of 1956, with particular reference to the National Smoke Abatement Society,[1] the only group concerned specifically with this issue.

Legislation for smoke abatement dates back to 1273, but despite this long history it cannot be said to have been very effective. Within the last century the story has been largely one of half-hearted legislative action and unheeded committee reports. Before 1956 the law on air pollution was determined by the Public Health Act (1875), the Public Health (Scotland) Act (1897), and the Public Health (Smoke Abatement) Act (1926). These provisions were fragmentary and by the 1950s they had been rendered obsolete by technological developments. Committee reports of 1946 and 1952 recommending further measures did not lead to legislative action. In 1954 the NSAS suggested that the law was so out-dated that the only sensible course was to scrap existing legislation and substitute a new and comprehensive instrument. In 1956 this aim was achieved, partly because of its activities, partly because of the effect on public opinion of the great fog of 1952, partly because of the report of the Beaver Committee, and partly because of initiative taken by Mr Gerald Nabarro, MP. It will be convenient to present narrative under these four headings.

## I The National Smoke Abatement Society

### (*a*) *Membership and finance*

Apart from the war years, the Society had made steady, if unspectacular, progress since its inception in 1929. After five years the annual income was still less than £1,000; but by 1950, £4,500 had been attained and more recently the figure has approached £10,000. The membership has similarly increased. Local authority members, for instance, numbered only 60 in 1932, but 500 had joined by 1958, including all the major authorities likely to be concerned with air pollution.

The members can be divided into three main classes: local authorities, commercial undertakings, and private individuals and societies. Many local authorities are concerned if for no other reason than that they are required to administer most of the provisions of clean air legislation. The commercial members are mainly industrial undertakings, both public and private, and organizations of them that are directly or indirectly interested in clean air. Manufacturers of smokeless fuels and organizations such as the Solid Smokeless Fuels Federation, have a direct interest, and critics of the NSAS have not been slow to point out that the Society is partially financed by members of this kind. Public bodies like the National Coal Board have also found membership worth while. Commercial members, though numbering only about one-third of the membership, contribute about 55 per cent of the money annually subscribed. Finally, there are what might be termed 'do-good' members, consisting of individuals and societies which are not concerned commercially with clean air. Examples of such members are the Scottish Council of Women Citizens' Associations and the Sheffield Council of Churches.

This diversity of membership, though a source of strength in some respects, can lead to internal dissension. In 1952, for instance, the National Coal Board attempted (unsuccessfully) to prevent the Society from issuing a report criticizing the use of 'nutty slack', which had been recommended by the Minister of Fuel and Power. More recently the Society has clashed with the Coal Utilization Council (another member) over the advertising campaign sponsored by the latter on behalf of the open coal-fire.

In passing, it may be noted that this analysis of the Society's membership suggests that the usual classification of pressure groups into sectional and cause (or 'promotional') groups is not entirely satisfactory.[2] The Society might be thought of as promotional, but this would obscure the sectional element in its composition. A more useful distinction would be one between groups that are the definite and appointed spokesmen for a particular section of the community, and groups that are primarily concerned with the propagation of attitudes (which may well be economically beneficial to certain sections of their support).

### (*b*) *Policy and tactics*

The general propaganda activities of the Society are severely limited by its small income. According to its journal, it is a question 'not of any blitzkreig [*sic*] but of an untiring war of attrition along the whole of the air pollution front'. 'It is . . . impracticable', the Society's report for 1957 states, 'to consider intensive or continuing publicity campaigns, and reliance has to be placed on steady, if unspectacular work.'

The Society has therefore usually been discriminating in the selection of targets. Generally, what the Society's 1948 report called 'the better informed sections of the community' have been chosen, on the view that this 'informed' minority would have to be converted before popular support could arise. References to this distinction between the informed and mass opinion occur frequently in the Society's publications. Reviewing the Final Report of the Government Committee on Air Pollution, *Smokeless Air* demanded that it be acted upon 'with a sense of urgency', and to secure this result, thought the journal, 'may well require the constant pressure of informed public opinion which can be secured not by costly mass propaganda, but rather by continued education and stimulation of the leaders of public opinion and those who will develop and administer the measures to be taken'. It should not be concluded, however, that mass opinion was totally neglected by the Society; for, as will be seen later, every effort was made to place items in the popular press. But even here there seems to have been more rejoicing over one item in *The Times* or the *Manchester Guardian* than over ninety and nine elsewhere.

Another main general principle of the Society's publicity tactics, which was indeed partially entailed by the first, was the determination to be at all times moderate and factual. The Society has shown 'a care that propaganda must be backed by sober data; an emphasis on realism in all advocated policies – all these have helped, and although with more hot air and glamour the Society might have made more of a stir from time to time, it would in the end have got it nowhere, and its prestige, instead of steadily rising, could only too easily have been lost'. The case for smoke abatement is, furthermore, one that lends itself readily to highly factual statement. There is evidence that the majority of people (even among the 'informed') have really very little idea of the economic and physiological toll of air pollution. Indeed, the Society itself was taken aback when the Beaver Committee in 1954 quoted the surprising figure of £250,000,000 as the annual cost of air pollution. Such lack of appreciation is understandable, for the deleterious effects of pollution are not, in normal times, readily apparent. The alarming facts were therefore expected to speak for themselves, and some of the Society's literature consists entirely of factual assertions that leave the reader to draw the right conclusions.

The Society's moderation can be seen clearly in its attitude to the *Interim Report of the Government Committee on Air Pollution* published in November 1953. This was a brief appraisal of the position together with some recommendations of emergency steps that might be taken in the event of the recurrence of a fog similar to that of December 1952 in London. The Report was not sufficiently radical for some sections of the press, and the Society could easily have joined in the clamour in an effort to secure a more vigorous Final Report. However, it chose to support the Committee and came down heavily on the critics: 'Any report short of a fairy story would have been a disappointment to the uninformed . . . it was the papers that had given little or no attention to the problem and hardly knew the first thing about it that were the most critical and downright silly.'

Above all, the Society had to avoid the crippling accusation of crankiness. In its early days it was often dismissed as an organization of eccentrics. Had this charge stuck, it would clearly have been damning to a body whose main hope was to gain a reputation for authoritative statements within its field.

The Society's authority and knowledge have been the main factors behind what success it has achieved since 1929. They have been the factors which have enabled it, in part at least, to overcome the disadvantages of its inability to stage mass campaigns.

### (*c*) *Media*

The propaganda used by the NSAS may be listed under five main headings:

*i Publications of the Society* The Society publishes a quarterly journal, *Smokeless Air*, which is sent to all members and to organizations such as libraries, and which now has a circulation of about 5,000. Apart from editorial comment, its contents can normally be divided into 'popular' material concerning the Society, its members, meetings, etc., and more or less technical material on the causes, effects, and prevention of air pollution. As well as being 'the most effective single means for maintaining the interest of the Society's members', the journal has been widely reviewed and quoted, particularly in the more specialized sections of the press which deal with fuel and industrial processes; and there are few issues which do not lead to press comment and summaries.

The Society's *Year Book* has a circulation of about 3,000 and purports to be something of a 'clean air handbook'. It certainly contains much information (bibliographies, summaries of recent official publications, etc.), that might reasonably be regarded as useful to persons interested in smoke abatement and allied topics.

The leaflets produced by the Society vary from general publications like *Facts about Smoke* and *Guilty Chimneys* to more specialized ones such as *Smoke Prevention and Science Teaching*.

*ii The press* Press publicity directly or indirectly attributable to the Society may be put in three categories. First, there is material contributed directly by members. In 1958 nine such articles appeared in various places. A single contribution of this kind may be of considerable importance: for example, the article 'Polluted air over towns' by Dr R. Lessing which appeared in *The Times* in 1953.

As well as such material there is publicity directly inspired, or at least greatly assisted by, the Society. The 1954 Annual Report found that 'many press articles have been published, frequently following interviews or correspondence with journalists and other writers', and the Society's literature contains frequent references to assistance rendered to writers of all descriptions. Indeed, the Society has increasingly come to regard itself as a sort of information bureau for inquiries about smoke abatement and allied problems: 'Journalists and other writers, technical workers, speakers, local authorities and others look to the Society to supply them with information or to put them in touch with other bodies, persons or sources that can help.' Government departments and nationalized industries have, apparently, been among the inquirers. Similarly, by giving journalists material 'ready to serve', the Society always receives good coverage for its annual conference; the Bournemouth Conference of 1955 occasioned no less than 467 items in various newspapers and periodicals.

*iii Lectures and meetings* The extent of the propaganda that can be disseminated in this way is clearly limited by the Society's lack of man-power. The Economic League could hold 18,000 meetings in

1956,[3] but at this time the NSAS had only three paid staff who gave lectures. However, oral propaganda on a limited scale, directed to carefully chosen audiences, has always been regarded as important.

*iv Exhibitions* Exhibition work is becoming increasingly important. Local authorities proposing to set up a smoke control area often hold a clean air exhibition, or devote some part of an exhibition to the clean air theme, and it is to the Society that they almost invariably turn for their exhibits. The Society's activities in this field have increased steadily, especially during the post-war period. In the year ending 30 June 1937 the Society participated in only four exhibitions, but in 1948 the corresponding figure was eighteen, and in 1957 the Society participated in over forty exhibitions of various kinds, while a further twenty requests for assistance had to be refused through lack of resources.

*v Broadcasting* Fog and smoke abatement have not received extensive treatment on radio and television. None the less, the Society's expertise has made possible a slight access to these most important media that would otherwise have been denied. In 1950, for instance, Dr G. Burnett (MOH for Preston and a member of the Society's Executive Council) gave a talk on air pollution on the North of England Home Service and in a television programme called 'Fog' Richard Dimbleby interviewed Arnold Marsh, General Secretary of the Society. More recently several members of the Society took part in a sound broadcast 'Science against smog' which was thought to have been very successful.

### (*d*) *Degree of success*

Clearly, smoke abatement is not a cause which may reasonably be expected to make spectacular advances. So long as many people regard pollution as almost a natural phenomenon education is bound to be slow. However, there can be no doubt about the rise in interest in and concern over air pollution. The growth in the Society's numbers and wealth might well be taken as evidence of this. Further, the press has paid ever-increasing attention in post-war years. Sixty newspaper cuttings a month on the subject were received by the Society in 1946, but by 1951 the figure had risen to nearly 200. By 1954 1,000 cuttings were arriving monthly, and the number has continued to increase. To some degree also, the extent of air pollution measurement may be taken as an index of the importance attached to the subject. In 1939 only 202 instruments were in use; by 1954 the number was 1,446.

More specifically, the Society has been responsible for the

propagation of the two ideas which have been the main vehicles of smoke abatement in recent years. It was in 1935 that Charles Gandy (Chairman of the Society) first suggested in a letter to the press that smokeless zones should be established in areas of high pollution. Although it was not until 1951 that the first smokeless zone was actually set up (in Coventry), almost all of our major cities now have at least one. Commenting on the establishment of the first zone, the *Medical Press* thought that 'the beginning of the experiment would still be some decades in the future' had it not been for the efforts of the Society. The principle of 'prior approval', whereby plans for the installation of fuel-burning appliances have to be submitted for the approval of the relevant local authority, was not invented by the Society. Indeed, it was in operation in several continental countries some years before the war; but the Society can claim to have made it familiar in technical and administrative circles in this country. The Clean Air Act of 1956 was largely based on these notions of smokeless zones and 'prior approval'.

The Society has the advantage that its cause is difficult to oppose openly, though indirect opposition is far from difficult because of the still strong romantic tradition surrounding the open coal-fire. It is perhaps indicative of the success of the Society's propaganda that interests which do not on the whole benefit from clean air policies have lately found it necessary to disseminate counter-propaganda. As early as April 1954, Mr J. Wood-Smith (vice-president of the Coal Merchants' Federation of Great Britain) was urging members to 'take up the fight to ensure that air pollution and smoke abatement are kept in their proper channels', and in 1955 Mr J. W. Stewart, presenting the annual report to the annual meeting of the Federation, said that there was a feeling in the trade that they were not 'standing up for the open fire quite as they should do'. Since the passing of the Clean Air Act, the coal merchants have taken the matter much more seriously. The Federation has acquired the services of a firm of public relations advisers (Sidney-Barton Ltd) and has actively supported a public campaign on behalf of the open fire sponsored by the Coal Utilization Council, on which expenditure of nearly £200,000 was apparently contemplated. It has also appointed special regional and national 'clean air officers', and has sought to persuade the Minister of Fuel and Power and the fuel and power committees of the Conservative and Labour parties of the justice of its cause. Locally the Federation has helped to represent objectors to proposals for the establishment of smoke control areas.

In general, it can be held that the Society succeeded in establishing a reputation for moderation and knowledgeability in its field. It numbered among its members experts on all aspects of the problem, and was widely looked to for information and assistance. The *Medical Press* summed up its contribution by asserting that while it was 'presumably true that the number of individual objectors to atmospheric pollution by smoke would have been increasing steadily . . . the legislative preparation of the ground could never have been accomplished by individuals. . . . Only an organized body of accepted standing could have produced the favourable social climate which was necessary.' Until 1952, however, the 'preparation of the ground' was all that had been achieved. Very few results had been secured. The narrative will now be taken up at the point of the great London fog of that year.

## II The great London fog

Despite the connexion between fuel efficiency and clean air which had been demonstrated many times before the great fog, not least in official reports, the Government had appeared far from eager to take any serious steps to prevent air pollution. Perhaps it was that, as Mr Gerald Nabarro later asserted, there were no votes to be gained from scouring the atmosphere. Twice during October 1952 Mr Edward Davies raised in Parliament the question of the adequacy of local authorities' powers. On neither occasion did Government spokesmen give the slightest indication that concrete measures would be forthcoming. Indeed, Mr Macmillan (Minister of Housing and Local Government) specifically rejected the suggestion that further legislation was necessary.

The great London fog lasted from 5 to 8 December 1952, and was afterwards said to have been responsible for 4,000 deaths. It seems likely that it was the worst fog disaster that has ever occurred; certainly it was the most publicized, although the immediate reaction of public opinion was 'strangely calm . . . almost fatalistic'. The newspaper cuttings received by the NSAS during the period support the contention of *Smokeless Air* that the fog had a 'relatively small effect . . . at the time it occurred'; and the aspect of the whole business which first attracted the headlines was undoubtedly the death or enforced destruction of several prize animals at the Smithfield Agricultural Show.

Several factors account for the absence of any immediate outcry. Air pollution does not produce its own peculiar disease; its effect is to exacerbate (albeit sometimes fatally) an already existing condition. Furthermore, the magnitude of the fog's assault on health was not immediately discernible. It was necessary to wait for statistical reports before even approximate information could be obtained. These causes, plus the frequency of serious fogs in London, partially account for the 'apathetic acceptance' of which the Society complained.

The Society immediately urged upon the Government a full inquiry, and later Arnold Marsh had an informal meeting with the medical advisers of the Ministry of Health and was assured that a full investigation was contemplated. The Society, however, was evidently not convinced that this was sufficient, and undertook its own survey, the purpose of which was to present the facts in a readily comprehensible and striking manner before memory of the event had faded. The survey was published in *Smokeless Air* (Spring 1953) and was sent to all MPs and members of the Government. Such was the subsequent demand that it was necessary to produce off-prints of the survey.

The evidence points to a considerable build up of public opinion, especially among the 'better informed sections', during 1953 which, directly or indirectly, led the Government to present at least an appearance of activity. The Government's first position was one of complete inaction.[4] On 17 and 18 December Government spokesmen refused demands by Mr Dodds and Mr Driberg for a special inquiry. Further questions followed in January and on the 27th Mr Dodds was less patient: 'Does the Minister not appreciate that last month, in Greater London alone, there were literally more people choked to death by air pollution than were killed on the roads of the whole country in 1952? Why is a public inquiry not being held?' In addition Mr Janner wanted to know if the Minister would 'consider the introduction of general legislation dealing with the subject'. Mr Macmillan was 'not satisfied that further general legislation [was] needed at present' and reminded his questioners of the difficulties that would be entailed by such legislation. For Mr Dodds this constituted an 'amazing display of apathy' and he promised to raise the matter again. Ten more parliamentary questions followed before 21 May 1953 and one evening paper quoted Mr Macleod (Minister of Health) as complaining that 'he seemed to get nothing except questions about the fog and its effects upon people's health'. On 8 May Mr Dodds, raising the matter on the adjournment, testified to the growing amount of public unrest: 'Since I became a Member of Parliament in 1945 I have certainly had a great deal of correspondence, but I have never had more correspondence on any subject than I have had on this question of air pollution.'

Just before this debate the Government had in fact announced its intention to conduct an inquiry and there can be no doubt that this reversal of policy was largely caused by the increasing alarm in many quarters.[5] The authors of the Annual Report of the NSAS had no doubts concerning the state of public opinion. They detected a 'rising tide', and implied that it was because of this that the Government had belatedly decided to appoint a committee of inquiry. The report cited those publications which had helped to mould 'informed opinion'. Among these were the Society's own survey, a detailed analysis of mortality during the fog which appeared in the *Lancet*, two articles in *The Times* by Dr R. Lessing, and a number of reports by medical officers of health in the London area containing prominent references to the effects of the fog.

A reading of the contemporary newspapers seems to show, in addition, that an article in the *British Medical Journal* at the beginning of January 1953, suggesting that the fog might have been responsible for 4,703 deaths, did much to establish the currency of this statistic. However, perhaps the most publicized document concerned directly with the fog was a report of the health committee of the LCC, which appeared at the end of January, and stated that in some respects the event had been more lethal than the worst outbreak of cholera in the nineteenth century.

Perhaps it would not be exaggerating to suggest that the fog rendered the situation 'unstructured', in that its effects were, in the long run, sufficiently sharp to make obsolete the frames of reference in terms of which people had been accustomed to 'understand' the phenomenon of fog. It is significant in this connexion that the use of the word 'smog' seems to have increased greatly during 1953. 'For smoke abaters, this word was a valuable addition to the political vocabulary; "fog" is almost a natural phenomenon, "smog" is an evil to be eliminated.' In January the *Daily Mirror* was still using 'fog' in its headlines, but by December even the *Daily Telegraph* was talking about 'smog'.

As the next winter approached, the Government seemed anxious to appear actively interested. On 13 November, for instance, Mr Macleod announced, amid what *The Economist* called disapprovingly 'a blare of publicity', that 'smog-masks' would shortly be available under the National Health Service, and when the Committee on Air Pollution[6] issued a short Interim Report Mr Macmillan was at pains to assure the House of Commons that everything possible was being done that might prevent the disastrous effects on health of another great fog.

## III The Beaver Committee

The Government Committee on Air Pollution was set up in May 1953. Its terms of reference were 'to examine the nature, causes and effects of air pollution, and the efficacy of present preventive measures; to consider what further preventive measures are practicable; and to make recommendations'. This would seem to indicate that although the Government was originally unenthusiastic, it did not attempt to prejudge the issue by restricting the terms of reference. Furthermore, the Committee was generally considered a strong one. The Chairman was Sir Hugh Beaver, who had a scientific training and was managing director of A. Guinness, Son & Co. Ltd. He was to become President of the FBI in 1958 and had had considerable experience on committees of inquiry. Sir Roger Duncalfe was Deputy Chairman. He was connected with the glue and chemicals industry and was chairman of the technical legislation committee of the FBI. The other members of the Committee were: Miss A. Boyd (Housing Manager for Rotherham County Borough); Dr J. L. Burn (MOH for Salford); F. W. Charles (an accountant); S. R. Dennison (an economist, formerly Chief Economic Assistant to the War Cabinet Secretariat); T. Ferguson (Professor of Social Medicine and Public Health, Glasgow University); Dr R. Lessing (consulting chemist and consulting engineer); Dr G. E. Foxwell (fuel technologist); G. Nonhebel (fuel technologist and Head of the Fuel Economy Section of ICI; C. J. Regan (Chemist in Chief, ICC); and Dr O. G. Sutton (Director of the Meteorological Office).

The members of the Committee could be classified in several ways, but perhaps the most significant classification for present purposes is that into 'interested parties' and 'experts'. The former were of two broad types: those connected with industry, and those associated with local authorities. Apart from the Chairman, the first category included Sir Roger Duncalfe, Mr Nonhebel, and Mr Charles. The members associated with local government were Dr Burn, Miss Boyd, and Mr Regan. The 'experts' were Professor Ferguson, Dr Foxwell, Mr Dennison, Dr Sutton, and Dr Lessing.

Three of the NSAS's leading members, Dr Burn, Mr Nonhebel, and Dr Lessing, were members of the Committee; and Dr Foxwell had written articles for *Smokeless Air* and was to become a member of the Society's Executive Council.

The Committee and its sub-committees held in all 133 meetings in addition to visiting a number of large cities and meeting the representatives of interested organizations.[7] The Committee's Final Report states that it took little formal evidence, 'rather taking the view that all interests were in agreement as to the objective and that therefore it was a matter for joint discussion to find the means'. This acceptance of the 'objective' as universally agreed was a considerable triumph for the cause of smoke abatement, for it laid down at the outset the framework within which the discussions took place.

Earlier committees which had examined the problem had mostly laboured in vain, but in 1953 there were grounds for expecting a better result. The effect of the London fog on public opinion had been considerable and the progress that had been made since the war in linking fuel efficiency with smoke abatement had led to an increasing realization that clean air could be good business. Finally, post-war technological improvements had made smoke abatement a much more feasible proposition, and administrative developments (for instance, the successful working of several smokeless zones) had not altogether failed to keep pace with progress in the scientific field.

The NSAS did not, of course, consider that the presence of its members on the Committee would be sufficient to ensure the friendly report for which it hoped. The Executive Council decided that a memorandum should be prepared containing a comprehensive summary of the Society's views. The result was a document of some 10,000 words which was submitted in January 1954. The memorandum was reprinted in the Society's journal and also published separately, and was reported or reviewed in over eighty newspapers and periodicals. Officers of the Society, its journal records, also had a 'preliminary informal discussion' with the Committee, which was further lobbied when visiting Manchester by members of the Society's North-Western Divisional Council.

A comparison of the Society's memorandum and the Committee's Final Report (published in November 1954) shows how closely, given minor deviations, the Committee followed the views of the NSAS in reaching its conclusions. The Society's report for 1954, welcoming the Committee's Report, commented that: 'In its analysis of the problem and the judgement made upon it, the report fully, and even vigorously, confirms the case as it has for many years been stated by the Society.' The document was received by the press with almost unanimous enthusiasm and seems to have been accepted (as the Society urged) as 'the most authoritative statement on the subject that can be expected for a considerable time to come and as the point of focus from which all future discussion and action will develop'.

Herein, perhaps, lay the fundamental significance of the Committee's work and Report. It was authoritative and provided a basis for future work. Local government, industry, and science had all been represented, and a unanimous Report had been produced condemning the evils of air pollution in an outspoken manner and recommending strict legislation to deal with the situation. Given the factors enumerated earlier that made the occasion conducive to the passing of such legislation, there seemed to be a reasonable chance of persuading the Government to act, if it was not already half-persuaded.[8]

## IV The Nabarro Bill and the Clean Air Act

### (a) *The Nabarro Bill*

A little before the publication of the Final Report of the Beaver

Committee, some dozen MPs, representing both sides of the House and including Mr Gerald Nabarro, provisionally agreed that if one of them was lucky in the ballot for Private Members' Bills, he would introduce a measure for smoke abatement. Mr Nabarro won first place and set about preparing a Bill. It has been seen how the Beaver Report was almost everywhere well received, and although the Government did not disclose its attitude until 25 January 1955, clearly Mr Nabarro's most propitious course would be to propose a measure that embodied some of the Committee's recommendations. At first, it seems to have been uncertain whether he would, in fact, propose clean air legislation. At this stage his doing so seems to have been contingent on receiving the assistance he wanted from the NSAS.

A few days before the meeting of the Executive Council for the Society on 16 December Mr Nabarro approached the officers for assistance. The Bill he had in mind apparently embodied all the legislative proposals of the Beaver Committee which were appropriate subject matter for a Private Member's Bill (i.e. those not directly involving Government expenditure). He was mainly in need of briefing, expert assistance with drafting, and, most important of all, money to meet the expenses involved.

The parliamentary time-table necessitated a rapid decision by the Society. Its officers urged Mr Nabarro to limit his proposed Bill to the authorization of general powers for local authorities to establish smokeless zones (this was considerably less than the scope of the measure that Mr Nabarro contemplated). However, they agreed to meet the costs of a parliamentary adviser. The 1955 report of the NSAS asserts that the introduction of the Nabarro Bill was 'contingent on him being given the support' of the Society, and as his Bill was given a first reading on 15 December it seems probable that he decided to proceed with the project after his first meeting with the Society's officers.[9]

The meeting of the Executive Council on 16 December confirmed the financial understanding that had been given, but no conclusion regarding the Bill's contents was reached. Shortly after this meeting Mr Marsh had further discussions with Mr Nabarro and Commander Christopher Powell (the parliamentary adviser). Mr Nabarro continued to reject the less ambitious Bill favoured by the officers, saying that the Government's attitude to such a truncated measure would be that it was hardly worth while and that action could be delayed until it had decided to move. Nevertheless, he was still anxious that the Society should be associated with all the provisions of his Bill and thought that the Government would be obliged either to take it over or to introduce its own comprehensive measure. For the time being, however, the Society maintained its highly equivocal position, and meanwhile Arnold Marsh prepared an extensive brief for Mr Nabarro's use during the second reading debate.

January 25th saw the first Government reaction to the Beaver Report. Mr Sandys (Minister of Housing and Local Government), replying to a question from Mr Nabarro, indicated that the Government was prepared to accept in principle the policy recommended. Neither was this all: 'I have already had preliminary discussions with representatives of local authorities and the Federation of British Industries. The first meetings indicate that industry would be ready to play its part in implementing this policy, and that local authorities would be prepared to accept the additional responsibilities involved.' In spite of these declarations, Mr Nabarro continued preparations for the second reading of his Bill. To increase publicity, he held a press conference and took the unusual step of stamping his parliamentary letters 'Gerald Nabarro's Clean Air (anti-smog) Bill Second Reading – Friday, February 4th, 1955'.

On 27 January he attended a meeting of the Society's Executive Council to explain the contents and purpose of his Bill. It seems that he was very persuasive, for at the end of the meeting a unanimous resolution was passed fully supporting the Bill and calling on the Government to adopt it or to substitute its own comprehensive measure.

The second reading went according to plan for Mr Nabarro. He stated at the outset that, given a full discussion and a Government promise to introduce its own Bill during the session, he would withdraw his Bill. Most speakers supported Mr Nabarro. Mr Sandys said that he did not resent Mr Nabarro 'pushing us on a little', but could not advise the House to accept the Bill. He maintained that it had been prepared without the benefit of consultation with the interested parties, contained drafting defects, and lacked the necessary financial provisions. He also rejected as being of doubtful constitutionality the suggestion that the Bill should be taken over by the Government at committee stage. However, he promised that, as soon as the Government's consultations were completed, it would introduce legislation on the lines recommended by the Beaver Committee.

What was the significance of the Nabarro Bill in the struggle? The Bill openly confronted the Government with a definite proposal. It is true that the Government accepted the Beaver Committee proposals on 25 January and that consultations with the interests concerned were in progress before that date; but it is difficult to say how far these activities were inspired by Mr Nabarro's determination to introduce his Bill. It seems unlikely that the Government contemplated no action, but there is some evidence that the Bill did at least serve to expedite events, although it may not have been quite the 'pistol at the Ministers' backs' envisaged by the *Manchester Guardian.* However, immediately before the second reading, the *Municipal Review* seemed in grave doubt concerning the fate of the Beaver Report, warning that 'in a government office, there is a nice, clean, empty pigeon-hole the same size and shape as the Beaver Report, and there are probably several nice, clean and tidy politicians who . . . would like to see this report suitably housed'. Later in the year, the *Review* again suggested that the Report had only escaped shelving because of Mr Nabarro. Sir Hugh Beaver himself thought that Mr Nabarro's action might well have been the decisive factor in persuading the Government to introduce the Clean Air Bill.

### (*b*) *The Clean Air Act*

The general election of May 1955 nullified the Government's promise to introduce a Bill within the session; but a Bill was in fact given a first reading on the day before the new Parliament rose for the summer recess.[10] The NSAS subjected the Bill to close scrutiny, and Mr Marsh sent notes to Mr Nabarro on its provisions for use in the second reading debate on 3 November. Unlike the Nabarro Bill, the Government measure encountered severe criticism. Mr R. E. Winterbottom (Labour) thought that it was 'a miserable, hypocritical shadow' of the Bill introduced by the Member for Kidderminster[11] and several members thought that they detected the influence of vested interests behind the provisions.[12]

Before the beginning of the committee stage on 9 February 1956, there was again close contact between the NSAS, Mr Nabarro, and Commander Powell (whom the Society had retained) in an effort to agree on amendments to be moved. These were agreed at a meeting on 12 December between Marsh, Nabarro, and Powell; and detailed criticisms of the Bill were sent to all members of the standing committee.

S. E. Finer and K. C. Wheare have pointed to the importance of the committee stage for pressure-group activity,[13] and in the present case such activity was clearly in evidence throughout all thirteen meetings. The general pattern was one of a sustained pressure from the spokesmen of groups which wished to have the provisions of the Bill made more strict that was resisted by the Minister (Mr Sandys) and his Parliamentary Secretary (Mr Enoch Powell). Sandys and Powell were occasionally supported by MPs from the Government back benches who broadly represented industrial interests. As a general rule, however, the Conservative Members of the committee were content to leave the defence of the Bill to their leaders.

The main groups pressing for tightening the Bill's provisions were the Association of Municipal Corporations and the NSAS. Mr D. Jones (who was a vice-president of the AMC), Mr A. Blenkinsop, and Mr R. E. Winterbottom seem to have acted as spokesmen for the first group (or at least to have supported its general policy). Mr Nabarro was, of course, the Society's main representative, but again Messrs Blenkinsop and Winterbottom moved some of the Society's amendments and spoke in support of others. Lieut.-Colonel Wentworth Schofield, a well-known figure in the cotton industry, and Squadron Leader A. E. Cooper, who was plainly acting as a spokesman for the chemical industry (he referred in the debates to 'those of us in the chemical industry'), were mainly responsible for the few speeches made in defence of industry. The Society had the satisfaction of seeing several of its amendments accepted, although the Government remained unmoved on the main contentious points of the Bill.

The Bill passed through its remaining parliamentary stages substantially unchanged, although Lord Milner moved further amendments desired by the Society when the Bill reached the Lords. The Society's report for 1956 called the Act 'the most important administrative step forward in the prevention of air pollution since the Public Health Act of 1875', and continued, 'the new law may not go as far as some would have liked, and some of its provisions may prove to be something short of the ideal, but such imperfections should not obscure its over-riding importance'.

## V Conclusion

Perhaps three conclusions follow from this survey of the politics of smoke abatement. First, the study shows something of what may be achieved by an economically feeble body in converting an apathetic, or even hostile, public opinion to its views. Thirty years ago visionary faith was required for adherence to the cause. Now smoke abatement is practical politics. Some reasons for this success have been suggested.

Second, Mr Nabarro's relationship to the Society deserves additional comment in so far as it fails to confirm the impression that may be gathered from the relevant literature of groups eagerly co-operating with MPs acting on their behalf in the House,[14] particularly with respect to the introduction of suitable Private Members' Bills. The usual picture is one of groups pressing in from the periphery towards the centres of political activity in order to achieve their ends, but in this case the MP was the initiator and the group was at first reluctant to assist his efforts.

Finally, the conclusion may emerge from the narrative of events after the London fog of December 1952 that securing the establishment of a committee of inquiry of some kind at a crucial juncture is a fundamental part of attitude group tactics. Such groups are consulted much less frequently by Governments contemplating legislation than are sectional-spokesman groups; and the setting up of a committee of inquiry at the right time can well be the most important factor in the achievement of their aims. When the Beaver Committee was first established, the NSAS commented that 'the Committee can hardly fail to produce a report that will be welcomed by the Society'. The converse of this proposition would be that there will often be occasions when groups will not welcome committees of inquiry; at the moment the position of the RSPCA with regard to vivisection and that of the Lord's Day Observance Society with regard to the state of the Sunday observance laws would appear to provide evidence for this assertion.

**Editors' note**

Since this article was written, a Private Member's Bill, introduced by Robert Maxwell, was passed by the House of Commons in 1968. The new Act contains 'useful provisions to secure further abatement of air pollution'. (See *Smokeless Air*, Winter 1968.) But for a suggestion that the impact of the Clean Air Act, 1956, has not been all that great and that the reduction in air pollution may be attributable to other causes, see H. A. Scarrow, 'The impact of domestic air pollution legislation', *British Journal of Political Science*, July 1972.

## Notes

1 In 1958 the Society became the National Society for Clean Air, but it will be referred to throughout by its old name, abbreviated to NSAS.

2 This classification is, perhaps, most clearly exemplified by S. E. Finer in an article in the *Twentieth Century* (October 1957): 'There are two main types of associations. There are some whose whole *raison d'être* is lobbying. These are "promotional" groups – the societies for improving this and pulling down that . . . . In the other type of association lobbying is only a sideline. Its reason for existence is to provide domestic services for its members, and to promote their interests. But these interests do from time to time necessitate an approach to government.'

3 S. E. Finer, *Anonymous Empire*, Pall Mall Press, 1958, p. 78.

4 This is meant to be a factual assertion. It is not implied that the 'inaction' was justified or otherwise.

5 Some people, it appears, were genuinely afraid of a recurrence of the fog. The *Lancet* (10 October 1953) stated that most Londoners had a 'wholesome dread' of such an event.

6 See below.

7 Of the 122 persons and organizations who communicated their views to the Committee (listed in Appendix XII of the Final Report) no less than 26 were members of the NSAS or connected with it in various ways.

8 The evidence regarding the Government's intention in setting up the Committee is confusing, but some of it (the strong membership, the acknowledged vigour of the chairman, and the wide terms of reference that it could easily exploit) at least points to the conclusion that the Beaver Committee was what K. C. Wheare would probably call a 'committee to camouflage' (see *Government by Committee*, Oxford University Press, 1955, pp. 89-92, for his classification of committees of inquiry), i.e. a committee appointed to prepare opinion as a prelude to government action.

9 In a letter to the Mayor of Twickenham (24 June 1955) Sir Ernest Smith (Treasurer of the NSAS) stated that when Mr Nabarro won the ballot, he had in mind two or three topics which he considered well-qualified for Private Member legislation, and that which ever one he chose would depend on the assistance he received from others.

10 In fact, according to the TUC Report for 1955 (p. 245), the Government produced proposed legislation based on the Beaver Report by March 1955.

11 *Hansard* (5th series), vol. 545, col. 253. Two parts of the Bill especially angered the critics: the three so-called 'escape clauses' whereby, it was alleged, unscrupulous industrialists might evade its main provisions; and the seven-year 'period of grace' which, it was thought, might unduly delay the operation of certain clauses of the Bill.

12 Mr Nabarro in his speech [*Hansard* (5th series), vol. 545, col. 1254] said of the Bill that 'the hand of the Federation of British Industries is writ large between the lines'.

13 See Finer's article 'In defence of pressure groups', *Listener*, 7 June 1956, and Wheare, op. cit., pp. 136-7.

14 For example, 'The views of any particular group will not appeal to one MP as readily as to another. It may be that a group will not be able to find anyone to express its views, but this is unlikely unless they run contrary to party principles. If party principles permit an MP to support a group, then it is almost certain that an MP can be found to represent it.' (J. D. Stewart, *British Pressure Groups*, Oxford University Press, 1958, p. 204.)

# Chapter 4

# Power lines across the South Downs


## H. R. Burroughes


> There exists in the South Downs an area of still unspoilt country . . . possessing great natural beauty and much open rambling land. . . . We recommend it unhesitatingly on its intrinsic merits as well as on the grounds of its accessibility.
>
> Report of the National Parks Committee (England and Wales) 1947, Cmd 7121, para. 39

With these words the Hobhouse Committee, set up in 1945 to advise the Government on a national parks policy, proposed much of the West Sussex administrative county for inclusion in a future national park. Although for various reasons the Sussex Downs national park has not materialized, the area was designated an Area of Outstanding Natural Beauty (AONB) in the 1960s. The county also houses an appreciable middle- and upper-class population, reflecting both the attractive aspect of the countryside and its proximity to London. Prosperous commuters and retired people have joined the established landowners to form an articulate and wealthy community. Into this area, with its attractive landscape and amenity-conscious social classes, the Central Electricity Generating Board (CEGB) sought to introduce in the early 1960s a major electricity transmission line. The resulting planning dispute was among the most bitter and protracted in the history of public electricity supply in Britain.

A wide range of interests may be involved in negotiating the route of an overhead electricity line. Besides local authorities and various official organizations which must be consulted, the proposed line may impinge upon the interests of statutory conservation authorities, public utilities, private developments, amenity groups and individual property owners, all of whom have opportunity for registering objections.[1] The Bolney-Lovedean 400 kV line[2] through

West Sussex aroused an unusual amount of concern both locally and nationally and contributed to the developing public interest in problems of conservation and land-use. While general interest in the case revolved around the issue of reconciling inevitable development with preservation of the countryside, local concern centred much more on attempts by individuals and groups to find alternative routes less injurious to their particular interests.

Our prime concern is to review the kinds of interests which were drawn into the dispute and the ways in which they sought to further their objectives. For these purposes the story falls conveniently into two parts. Before the Public Inquiry vociferous protest groups were concerned primarily with defending private interests, and much activity was aimed at winning the support of the various local authorities for particular viewpoints. Most of the excitement occurred in the eastern part of the proposed route. Following the Inquiry, amenity groups emerged to play a more vigorous part, their attention being focused largely on the western part where the proposed route gradually converged with the South Downs. While the active parties were still largely private, the issue was perhaps nearer to a dispute about the public interest. Their immediate aim differed somewhat as it was more directly channelled to influencing the Ministry. This division is somewhat arbitrary as both types of issue and all sections of the line continued to be discussed throughout, and, of course, most participants attempted to present their particular arguments in terms of a wider public interest. Other participants also expressed views, but in the main these were more limited representations not creating such excitement and will therefore be omitted. The negotiations between the local planning authority and the CEGB, however, must be surveyed if the more local political conflicts are to be fully understood.

## The case for the line

The CEGB required a major line running from east to west through West Sussex on three counts. There was the need to reinforce supplies to the central Sussex area and to south-east England as there would be insufficient generating capacity to meet needs by the mid-1960s. On a larger, regional scale the line would be part of the high capacity 'Home Counties Ring', an irregular square around Greater London providing the conurbation's import requirements. Finally, on a national scale, it would form part of the main transmission line from Kent to Cornwall reinforcing the whole South Coast and providing outlets for big new power stations, a vital part of the National Grid system. Clearly it was impossible to refute the case for the line, the issue being merely one of how and where. As the cost of underground cabling for a line forty miles long would be prohibitive, the issue was essentially one of deciding the least damaging route.

This project began in a practical way during 1959, although it must have been foreseen at an earlier date when the Dungeness nuclear power station scheme with a line westward was prepared. However, no consideration was given actually to routing the Sussex line until the latter part of 1959.

The western termination was fixed before route-finding began as Lovedean substation, about six miles north of Portsmouth, already existed. At the eastern end, the substation site was not selected until 1961, but the general location was already determined by the route chosen for the Lydd-Bolney section in consultation with the East Sussex County Council. In that section the Board had been obliged to keep well inland for amenity considerations and to meet the wishes of the East Sussex planning authority, although there would have been technical advantages in following the coast. Thus when planning began in West Sussex the most direct route was to the north of the South Downs. Had strong arguments been advanced for taking this section by way of the coast it would still have been possible. However, the Board had made its own survey of such a route and had concluded that 'any such line could only be a major affront to the amenities' of the area. No responsible body sought to dispute this view.

## The County Council's dilemma

The Board began consultations with outside bodies in April 1960 with an informal approach to the West Sussex County Council. A meeting was proposed to discuss the siting of a substation in mid-Sussex (eventually Bolney) and the continuation of the line westward. The Clerk to the County Council preferred discussions to be at the officer level at that stage, and Mr S. H. Baker, the Deputy County Planning Officer, was responsible for conferring with the CEGB. During early summer Baker accompanied the Board's wayleave officers when they were working out the route on the ground, and there was the usual give and take. The Board's staff had Baker's tentative views by late July 1960.

Until the spring of 1961 the County Council was more concerned with the siting of the Bolney substation because of its importance for consequent line routes. The Board's plans for an overhead line were reported to the full County Council in February 1961, and obtained brief mention in the local press. At this meeting, however, it was the substation which took the Council's time. In August the Board announced that it had bought land in the Twineham

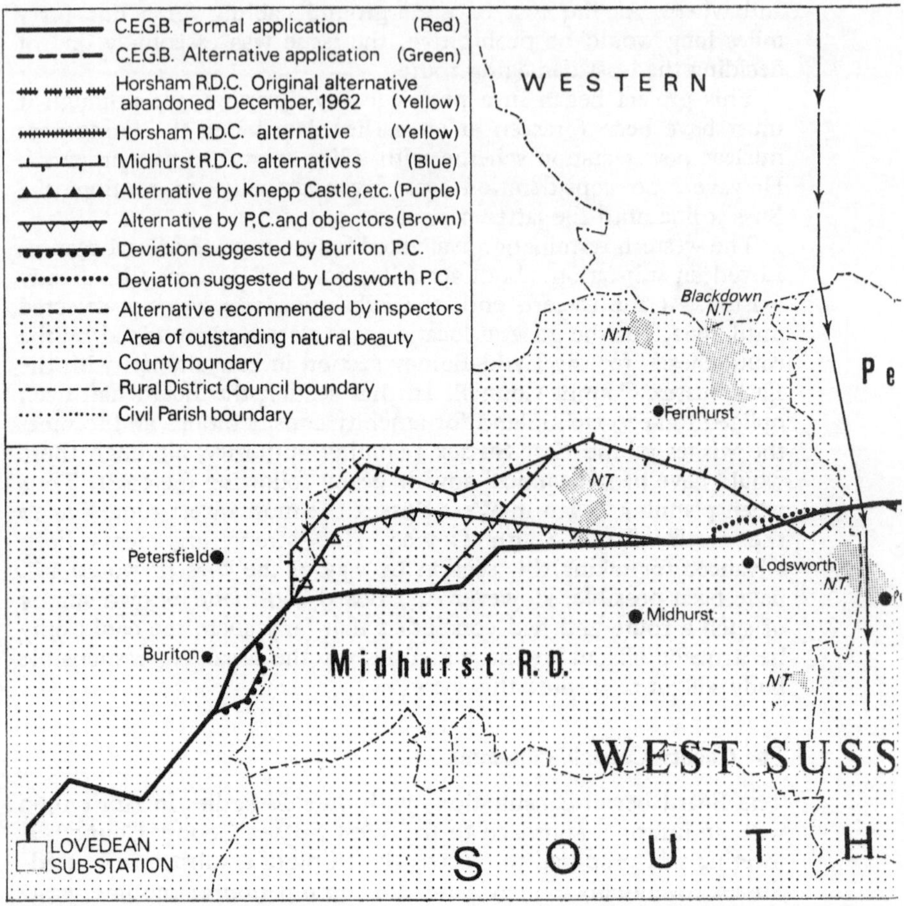


**Bolney–Lovedean proposed routes of 400 kV electricity transmission line.**

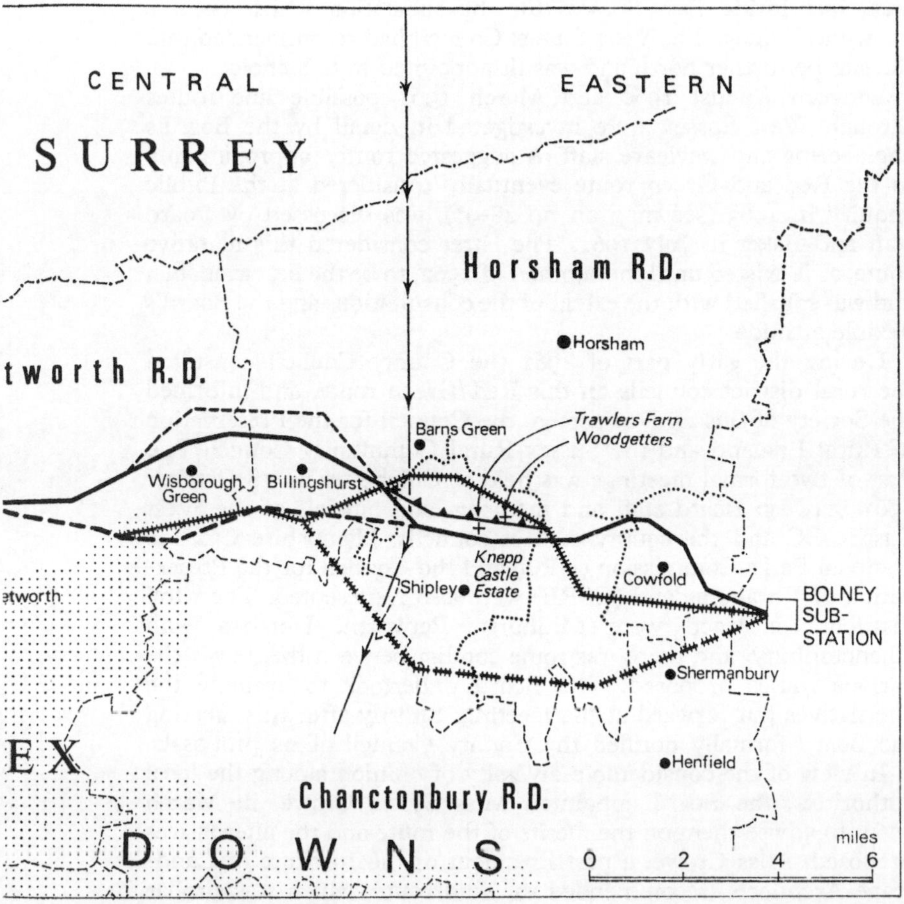
CENTRAL
EASTERN
SURREY
Horsham R.D.
Horsham
tworth R.D.
Trawlers Farm
Woodgetters
Barns Green
Wisborough Green
Billingshurst
Knepp Castle Estate
Cowfold
etworth
Shipley
BOLNEY SUB-STATION
Shermanbury
EX
Henfield
Chanctonbury R.D.
DOWNS
miles
0
2
4
6

area, just inside East Sussex, for the substation which became known as Bolney. The West Sussex Council had recommended that the site be further north and was disappointed in this choice.

Between August 1960 and March 1961 possible line routes through West Sussex were investigated in detail by the Board's engineering and wayleave staff. A suggested route, approximating to the Red and Green route eventually considered at the Public Inquiry in 1963 (see map on pp 48-9*), was discussed by Board staff and Baker in July 1961. The latter considered this tentative route, as it existed until the summer of 1962, to be the best available, and was satisfied with the extent of the consultations and the Board's flexible attitude.

During the early part of 1961 the County Council consulted the rural district councils on this Red/Green route, and informed the Society of Sussex Downsmen, the Council for the Preservation of Rural England and the Sussex Rural Community Council.The first of two formal meetings was held at Chichester on 18 October 1961 between Board staff and members and officials of the West Sussex CC and the county district councils. Hampshire CC, the National Parks Commission (NPC) and the Council for the Preservation of Rural England (CPRE) were also represented. The rural districts concerned were Midhurst, Petworth, Horsham and Chanctonbury, and there was some conflict between them over the various routes proposed. The Board undertook to examine the alternatives put forward at the meeting. Shortly after this meeting the Board formally notified the County Council of its proposals.

In view of the considerable diversity of opinion among the local authorities, the Board appointed Miss Sylvia Crowe, in March 1962, to advise them on the merits of the route and the alternatives proposed. Miss Crowe, a past President of the Institute of Landscape Architects, recommended the Red route, which was similar to the Board's own route except in the Central Section where it had favoured the Green alternative further south.

A second formal meeting was held at Chichester on 20 September 1962, with the same bodies again represented. The County Council asked the Board to consider the views expressed and to submit a formal planning application. At the meeting the Red route was discussed and there was much disagreement between the local authorities, although in the Central Section both the RDCs and the County Council were united in their dislike of Miss Crowe's route. However, it appears that the County Council did not decide even on this section until a later date. The Board considered the

* The various routes discussed are shown by different markings on the map. In the narrative, however, it is more convenient to retain the colour references which were used during the Public Inquiry.

Chichester discussions and Miss Crowe's advice, and on 1 October decided to make formal application to the Minister of Power for his consent to the Red route.

In deciding its attitude the County Council had the advice of its own staff, as well as the views of the district councils, various of whom sought the county's backing for their own alternatives. Besides the observations of those bodies which it approached, most of the local preservation societies wrote to the County Council expressing their views. As early as August 1961 the County Council had begun to receive letters from individuals and parish councils.

Realistically, the County Council accepted the line as inevitable and also accepted that the Minister of Power would not agree to underground cabling because of the high cost. Only the precise route of the line was open to debate. Eventually the County Council decided the original route for the Central Section was preferable, with the line south of the villages of Billingshurst and Wisborough Green, and announced in January 1963 that the Board had agreed to make formal application for the Green alternative. Thus no objection would be made to the Green route but opposition to the Red route there would be maintained. Still no final decision had been reached over the Eastern and Western Sections. As a result of local criticism about its non-committal attitude, the County Council felt obliged to issue a statement explaining its position. A formal objection to the line had been lodged with the Minister of Power to ensure that an inquiry covering the whole route in West Sussex would be held.[3] While the Planning Committee objected to the entire route, they accepted that a line was needed, and on the evidence available felt that only in the Central Section could they suggest a route demonstrably less objectionable. Counsel had been instructed to present the Council's case at the Public Inquiry scheduled for March and to help the Inspectors in their search for the least harmful route. When the Planning Committee reported to the full Council on 22 February no member made any comment. By supporting the Green route in the Central Section the Council was adopting the views of its planning staff. In the Eastern and Western Sections the Board route was thought least objectionable by the Deputy Planning Officer, but here the Council offered no evidence and voiced no opinion. In the circumstances, with widespread agitation and local opinion deeply divided, the County Council could hardly have been expected to do more.

## The politics of local planning

The uneasy posture of the County Council becomes more explicable when the attitudes and actions of the district and parish councils and local residents are considered. With more limited areas and lacking broad planning responsibilities, rural district councils are more likely to respond to immediate pressures of local interests. They do have a statutory role, however, and it is worth while for interested parties and pressure groups to seek RDC support. In the Eastern Section there was competition for the official support of the Horsham RDC. However, as events in the other areas were markedly less dramatic these can best be dealt with first.

### *Western Section*

During 1961 Midhurst RDC was invited by the County Council to take part in discussion of the Board's tentative route, and it informed parish councils of the route. Lurgashall Parish Council raised the possibility of the line following the Pulborough-Petersfield railway, in the centre of the Rother valley and closer to the South Downs than anything proposed by the Board. At the first Chichester meeting the RDC asked the Board to investigate this route, but subsequently emphasized that this suggestion was put forward on behalf of a parish council and not in any way endorsed by the RDC.

After the second Chichester meeting in 1962 the RDC again examined the problem, and the two Blue alternatives were put forward. Told of these routes, the County Council adopted a neutral attitude. During December the Board refused a request to advertise the RDC's alternatives, and declined to make formal application for them. The RDC did not specifically advertise these routes as, formally, the Inquiry was not about them, but there was publicity in the press, and they were the subject of letters written by the RDC to the parish councils and amenity societies. At the Inquiry Midhurst RDC presented evidence in favour of the Blue routes and against the Red. Three local amenity societies, the Society of Sussex Downsmen, the Sussex Vigilants Association and the Harting Society, were aligned with the RDC while the CPRE and the National Trust argued the converse. Other amenity societies, two parish councils and a score or so individuals were also represented, arguing for or against various permutations.

### *Central Section*

Of the two RDCs affected, Petworth decided in November 1962 to favour the Green route, and this attitude was adopted by the parish councils of Wisborough Green and Kirdford. Horsham RDC had always been convinced that the route should go south of the villages of Billingshurst and Wisborough Green. The landscape consultant retained by the RDC (see below, p. 54) reported in favour of this, his Yellow route differing only in minor detail from the Green route. Billingshurst Parish Council also expressed support for such a route. At the Inquiry the CPRE joined the County Council and Horsham RDC in arguing for the Green route as against the Red, while the National Parks Commission, the Nature Conservancy (worried about the effect of tree-felling on beetles) and the Sussex Naturalists' Trust Ltd made contrary representations.

The lack of parish politics in this area is rather surprising as the villages of Wisborough Green and Billingshurst were each bitterly divided over the issue. A developing dispute between property owners north and south of Wisborough Green was reported in November 1962, the change to the Board's route being attributed to protests from those affected by the original line. A petition against the Red route, organized by an 'Action Group' living in the Wisborough Green area, was sent to both the Petworth and Horsham RDCs during November. By January 1963 Wisborough Green was said to be completely divided, with the two groups of disputants as mutually hostile to each other as to the Board. The truth of this is seen in the report that in September 1964, a year and a quarter after the Inquiry, the scars had still not healed.[4] A group of forty-six objectors to the Red route went to the considerable expense of retaining a landscape architect and being represented by a barrister at the Inquiry. Interesting evidence produced by one objector was a set of photographs showing the visibility of proposed line towers, based on flying a balloon at the necessary height. Private objectors also appeared against the Green route, counsel being retained in two instances. Altogether a lot of time, energy and money went into the dispute round these two villages, but for the really interesting political activity we turn to the final area.

### *Eastern Section*

Horsham RDC was one of the first district authorities to give serious consideration to the proposed line. A route well to the south of later suggestions was favourably considered by the Council's Planning Committee in July 1961. At the same time a pylon sub-committee was appointed with Lady Burrell, Chairman of the Planning Committee, as one of its members. This point was later seized upon by opponents as she was, throughout the controversy, a director of the Knepp Castle Estate Company, whose land is crossed by the Red route. This sub-committee, set up to investigate a less objectionable route westward from Bolney, reported to a special RDC meeting in August. Subsequently Horsham put forward this route, running south-west through Shermanbury to within a few miles of the South Downs, at the first Chichester meeting, but was inevitably opposed by Chanctonbury RDC into whose area the line was thus moved. Horsham disapproved of alternative proposals put forward by the Board for the line to go north of Cowfold, and considered this to be much worse than the Board's earlier suggestions that it would go more directly westwards from Bolney.

A year later, in the autumn of 1962, Horsham still preferred a route following a south-westerly curve from Bolney to Wisborough Green. At the November Council meeting the Chairman of the Planning Committee reported that such a route was better than either of the two routes then being suggested by the Board because of the small population in that area. However, the County Council had already made clear its opposition to such a southerly route. After receiving a number of objections, including a petition against the Board's proposals, the RDC took the unusual step of engaging a consultant landscape architect, Mr Milner White, Vice-President of the Institute of Landscape Architects, to advise on the three suggested routes. Milner White made his preliminary report to the sub-committee of the Horsham Planning Committee on 3 December 1962, and on his advice the far-south route was abandoned, his view being that the Board's original route was the least objectionable. The RDC was somewhat dismayed by his report, Lady Burrell having emphasized the desirability of this south route at the Council meeting a few days previously. Chanctonbury, of course, welcomed the removal of the line entirely from its area.

At this time a route to the immediate south of Cowfold was not being considered but, following press reports that the Board's route was now favoured, the RDC received letters of objection, including one from a group of substantial landowners to the north of Cowfold enclosing a plan showing a possible alternative (which coincided with the final Yellow line in the Cowfold area). The RDC Committee resolved that Milner White should investigate this; however, he interpreted his instructions more broadly, and considered other alternatives before coming to his conclusions. He made his report to the Planning Committee on 15 January 1963, and subsequently the RDC put forward the Yellow line advocated in this report. The County Council agreed to support the RDC's request that the Board formally publish this alternative route, but would not support the route itself, thus provoking dissatisfaction among some of the Horsham councillors. The RDC was afterwards informed that the Board was only applying for the Red and Green routes.

When Horsham RDC received the Milner White report it had also for its guidance resolutions of the Cowfold and Shipley Parish Councils opposing the Red route and favouring something akin to the Yellow, and two considerable petitions of similar persuasion. These served to reinforce the views expressed in Mr Milner White's report. The Planning Committee meeting of 15 January, chaired by Lady Burrell, approved the Yellow route unanimously,[5] and it was also unanimously adopted by the full RDC on 25 January. Following the publication of the Yellow route the RDC received many letters, nearly all opposed to their alternative. Petitions against the Yellow route, totalling 1,100 names, were received and support on the parish councils began to crumble. A committee of the Horsham RDC apparently considered the matter once again on 27 February in closed session, two days before the final date they had fixed for receiving representations,[6] and reaffirmed the Council's decision to present the Yellow route at the Inquiry. Again, on 6 March the Planning Committee decided, at a meeting to which all other councillors were invited, to continue to support the Yellow route, although by then one parish council had withdrawn its support.

From the beginning of 1963 the action had increasingly moved to the parishes, and it is now time to turn from the RDC to the two parishes most immediately affected. To simplify the narrative this Eastern Section will be considered in two parts: in the eastern part the focus of activity was principally Cowfold parish, where the Yellow line goes to the south of both the village and the Red line; while in the western part, where the Yellow line is to the north of the Red, events centred mainly on Shipley Parish Council.

Shipley Parish Council had discussed the proposed line as early as September 1961 when the two routes in the north of the parish were considered. One of these approximated to what finally became the Yellow route in this area, and the meeting decided in its favour. In November 1961 Shipley again considered the problem and decided to commend to the RDC the route which was similar to the later Yellow route. At the same time it too set up a pylon subcommittee with Lady Burrell as one of its members.[7]

Following the Board's formal application Shipley Parish Council voted in December 1962 against the Red route, and during the month an eighty-signature petition was sent from the area both to the RDC and to the Minister of Power protesting about the Red route. Following publication of the Yellow route by the RDC, the Parish Council reiterated its support on 18 February 1963. At this meeting one of the members reported that there was public disquiet both about the Council's standpoint and the way this had been reached. However, allegations of secrecy were denied, one member commenting that he had talked with as many people as possible. The Parish Council decided by eight votes, with one abstention, to support the RDC's line 'as it is so near our line we proposed in 1961'.[8] It was further decided to write to the RDC informing them that should the Yellow route be rejected, Shipley had worked out an alternative route, and to ask whether the RDC would press for this on their behalf.

However, critical opinion locally was already finding expression, and a letter commending the County Council policy in this area found favour with one of the local newspapers, which added the heading 'logical, practical and wise'.[9] The writer condemned the Yellow route, weaned and reared in Shipley parish for ultimate presentation to the RDC prior to expert endorsement and adoption. He returned to the attack later in the month, accusing Horsham RDC of giving more attention to people to the south than 'to the far more numerous, though possibly less powerful, inhabitants north of the alternative line . . .'.[10] This correspondent collected over 400 signatures from the neighbouring Southwater area, while another opponent collected about 200 signatures from the Brookes Green, Barns Green and Itchingfield area; these petitions being sent to the RDC at the end of February.

Postponement of the Inquiry for a month, due to the illness of Miss Sylvia Crowe, allowed more time for these local wrangles to develop. The annual parish meeting took place in Shipley on 4 March attended by a number of dissatisfied parishioners from the north. Reminding the Chairman of their conversation about allegations of secrecy, one resident observed that the discussion had occurred only four days earlier when no mention had been made of either the recent Parish Council meeting or the forthcoming Parish Meeting. The Chairman apologized for this omission. Another opponent spoke of the adverse effect on over 100 houses in the Southwater area, and observed that the proposed alternative would be beneficial for only a very few people out of the total population of the parish. Another local resident described how Horsham Planning Committee had asked for changes in the plan and materials of his bungalow because appearance was important as it was sited on the highest ground in the neighbourhood. He now found the Yellow route passed over his bungalow and that a 160-foot tower would be placed in the same field! How, he asked, could the Planning Committee reconcile these two attitudes. After further criticism, a resolution was carried by twenty-five votes to four instructing the Parish Council 'to make urgent representations to the Horsham Rural Council asking them to withdraw their objection to the CEGB line and to support that line'.[11] At its meeting immediately afterwards, however, the Parish Council decided that, while it would forward this resolution, the accompanying letter would reaffirm its support for the RDC route.

Yet another alternative, to run between the Yellow and Red routes, was introduced by Lady Burrell at a Parish Council meeting at Shipley on 29 March 1963. After she had withdrawn from the meeting the Council decided to inform the RDC that they would support them should they adopt such a route. This route, it appears, was similar to the Purple route proposed later at the Inquiry on behalf of the Knepp Castle Estate Company. An explanation of this move advanced at the Inquiry was that until the extent of public opposition became known in February 1963, the Knepp Castle Estate Company was relying on the Shipley route of November 1961 (approximating to the Yellow route). This estate occupies a large part of Shipley parish,[12] and some people saw the Yellow and Purple routes as attempts to remove the line as far from the centre of the estate as possible, the latter not actually moving it off the estate.[13] The Purple route itself was introduced at the Inquiry as part of the case of a group of objectors, including the Knepp Castle Estate Company. Counsel disclosed that the line was drawn by Sir Walter Burrell.[14]

At the Inquiry Mr H. C. N. Henry, who lived at Woodgetters in the north of the parish, contended that Horsham RDC had no *locus standi* to support the Yellow route during the proceedings. His argument was the part Lady Burrell had taken in the decision to put the Yellow route forward. Lady Burrell was stated to be Chairman of the RDC Planning Committee as well as a member of the full Council, on the special sub-committee set up by it to study the matter, and on the Shipley Parish Council and its pylon sub-committee. The 1961 Shipley route had been recommended by the Parish Council to the RDC, with Lady Burrell a member of both the relevant committees. At the Inquiry Henry produced evidence to show that no member of Horsham RDC had sought dispensation from disabilities imposed by the 1933 Local Government Act from the Minister of Housing and Local Government, or any Shipley parish councillor a similar dispensation from the County Council.[15] Although the Knepp Castle Estate Company was represented at the Inquiry by both a QC and junior counsel, the offer by the inspectors to permit Henry's cross-examination at a time convenient to the QC was not accepted. The examination by the junior counsel did not resolve the issue of whether Lady Burrell's interests in the Knepp Castle Estate were such as to incur any disqualification, but Henry was on unfirm ground in suggesting that any infringement would necessarily have invalidated Horsham RDC's participation in the Inquiry. He had Knepp Castle Estate land on three sides of his property, yet despite the Company's assurances at the Inquiry that all affected by the Purple route had been notified, he maintained he had been told nothing. He also pointed out that the Purple line, after passing close to his property, turned north to avoid Trawlers Farm, where a Burrell relative lived. He further implied misrepresentation in the gaining of the signatures of two landowners in support of the Purple route, reading at the Inquiry letters from them objecting to this route, and was reluctant to accept the Knepp Castle counsel's suggestion that this resulted merely from misunderstandings.[16]

Neither was Henry alone when he refuted the Knepp Castle contention that the effect of the Red route on the proposed new Burrell Arms at The Bar, on the A24 road, was a forceful argument against that route. He was reported as saying, with characteristic bluntness, that they wanted 'the line shifted half a mile up the road to avoid their non-existent pub, whilst at the same time they run their Purple line right across the full length of my southern boundary, within 100 yards of my house'.[17] Counsel for a group of objectors argued that it seemed a little singular that the question of the building of the new public house was put forward as an argument against the Red route. Those who had applied for the permission and those who granted it must have known perfectly well that the site was on the route proposed by the CEGB a year previously. Presumably the application had been made to the Horsham RDC which had Lady Burrell as its Planning Committee Chairman, therefore the effect of the Red route on the site could not have escaped the notice of either the applicant or the committee.[18] This reinforced further the lack of confidence in the stand which the Horsham RDC had taken, which might be summed up by saying that it appeared, in the phrase used by one local newspaper, as being 'designed to please the big estates?'[19]

Turning to the neighbouring Cowfold parish, at the eastern extremity of the line, there was further controversy about the RDC's route, and during the final months before the Inquiry there was busy local political activity. Besides Horsham, another RDC had an interest here. While the CEGB proposals did not affect Chanctonbury, this RDC had long been apprehensive about alternatives put forward by Horsham, and Chanctonbury councillors were much relieved in December 1962 to hear that the far south route had been abandoned. Its Chairman, Mr S. W. Fowler, was told on 7 December by Horsham's Chairman that no further routes south of Cowfold were contemplated.

The Cowfold Parish Council meeting of 14 January 1963 received news of the latest proposal for a route south of Cowfold, and at this time 160 people in the area had petitioned Horsham RDC in favour of such a route instead of the Board's Red line. By five votes to two, with one abstention, Cowfold found against the Red route and in favour of a southerly alternative. Following receipt of the official RDC proposal the Cowfold Chairman called a special meeting for 14 February. Before this date Fowler sought out the feeling of Cowfold residents, and was instrumental in collecting a petition, although none of the ten canvassers lived in Cowfold itself. Fowler, who had considerable business and family interests in Cowfold, declared that one of the activating influences on him was that RDC and County Council members should disclose their interests.[20]

At the time of the Cowfold Parish Meeting on 14 February there were 470 names on the petition requesting the Council to ask the RDC to withdraw its Yellow route, most of those who signed living within the parish. There is evidence that much interest had been aroused, with fifty people attending the February meeting and sharp comment being made about who really was promoting the Yellow route. Fowler suggested that the 160-name petition against the Red route was mainly signed by tenants of northern landowners. Correspondence received from Mr Fowler was one of the principal causes of the 14 February meeting. Mr Moore, the Parish Council Chairman, regretted that he had not been told of the petition earlier so that he could have prevented recent events. Subsequently he had carefully sounded out public opinion in Cowfold and had concluded that the people were both very concerned about having their skyline blotted out and 'sick and tired of what is left in this place of the feudal system' and that 'they want no more of it'. Opponents of the Red route queried whether the previous decision could be reversed, but Moore was adamant that public opinion must be respected and commented that 'this thing goes deeper than what is thought'. By four votes to three with one abstention the Parish Council, in response to public pressure, reversed their previous decision and agreed to request the RDC to withdraw the Yellow route.[21]

Thus, during the final few weeks before the postponed Inquiry began, the pattern of support within the Horsham area changed radically. The RDC's position had declined through disaffection and had been undermined by the emergence of considerable support for the CEGB route. Only one of the two parish councils directly affected still supported Horsham, and even here the annual parish meeting had repudiated the route. The County Council was non-committal but had commented that the route was longer than the Board's. Chanctonbury RDC had declared its opposition and was to be represented by counsel at the Inquiry, and the neighbouring parishes of Shermanbury and Henfield had also made their protests. Already both the CPRE and the Society of Sussex Downsmen had come out in favour of the Red route, and at the Inquiry the official CEGB route was opposed only by Horsham RDC and one group of private objectors. Allegations, not adequately refuted at the Inquiry, that the boundary between private and public interests had not been precisely defined cannot have helped further the cases of these objectors.

**The Inquiry**

The postponed Inquiry began at Chichester on 2 April 1963 and turned out to be a monumental affair, the hearing of evidence taking nineteen days extending over six weeks – the longest electricity planning inquiry to date. Presiding over the proceedings were the Chief Engineering Inspector of the Ministry of Power and a Senior Planning Inspector from Housing and Local Government. As usual the Inspectors played a rather passive role,[22] but with fifteen barristers (including four QCs) and a multitude of other participants few aspects remained unexplored. Formally the Inquiry was concerned only with the Red and Green routes, for which the CEGB had made application for ministerial consent. Actually all definite proposals made by interested parties were investigated, including one or two alternatives which first appeared during the course of the Inquiry. However, the Minister can only consent to routes sought officially and, should another route appear advantageous to him, a new CEGB application is required with, perhaps, another inquiry. Much frustration occurs when it is forgotten that such inquiries are only held to assist a Minister to arrive at the best possible administrative decision and to allow aggrieved parties to make representations.[23] Although inspectors make recommendations, the decision is the Minister's.

Presentation of the Board's case took three days, including the evidence of its landscape consultant, after which other interested parties, from County Council to solitary private objector, had their say.[24] As is usual at such inquiries, most time was spent in considering the advantages and disadvantages of the various options in terms of their effect on landscape, private property, economic interests and areas of scientific interest. Landscape architects, surveyors, amenity society representatives and local residents testified on the differing capacities of various areas to absorb this line with its twenty-four heavy conductors carried on 160-feet-high towers. Witnesses for amenity societies invariably showed an extensive knowledge of the locality and a predilection for enumerating large numbers of views from footpaths, a game expert witnesses could not always play. The dispute about the legal standing of Horsham RDC was rather unusual, as also were the great expenditure on professional representation and witnesses by private individuals and the extensive nature of many of the cases presented.

### Awaiting the decision

Following an inquiry there is normally a tailing off of interest while the inspectors conclude their investigations, write and submit their report and the decision is formulated within the Ministry. Most public and local authorities regard the matter as *sub judice* and refrain from further activity, as in this case. However, not all participants were so resigned. Interest now shifted to the Western Section of the proposed route, from Petworth to the Hampshire border. Whilst the motivation underlying the conflict in the east had been largely one of concern for private property, in the west interest was primarily with landscape amenity and the participants were mainly amenity societies. This is not to suggest that the amenity societies were uninterested in the Eastern and Central Sections; far from it, the Society of Sussex Downsmen, for example, gave evidence on both sections. With but one exception, however, there was agreement amongst the amenity interests that if the line was inevitable the Board's Red route in those sections was the least objectionable.

Towards the west the Red route converges with the South Downs, and finally, between Petworth and the Hampshire border, traverses part of the South Downs Area of Outstanding Natural Beauty (then awaiting formal designation). At the Inquiry amenity societies had made conflicting representations about this Western Section. Both the CPRE and the National Trust supported the Red route, while the statutory National Parks Commission considered it the least damaging. This view was opposed by the Society of Sussex Downsmen, the Sussex Vigilants Association and the Harting Preservation Society. Some opponents of the Red route felt after the Inquiry that the case against it had not been put with sufficient strength and the Society of Sussex Downsmen took the initiative in setting up an Action Committee. At that stage the aim was to appeal to the Minister to consider underground cabling for the Petworth to Petersfield section. The Downsmen approached both national and local amenity organizations, and ten locally based organizations appointed representatives to the Action Committee.[25]

Formed in August 1963, the Action Committee immediately launched a very active campaign. In October a leaflet headed 'The pylon protest' was issued, setting out the case for underground cabling. About 20,000 copies of this leaflet were printed and distributed in such ways as door-to-door delivery, being sent out with rates notices, deposited at local public libraries and delivered with daily newspapers. People interested were urged to write to their MPs and the MP for Horsham reported receiving over 1,000 letters. Many other MPs were approached and the Downsmen record their indebtedness to a number for the interest and time they devoted to the issue.

Much publicity in local newspapers resulted, with some mention in the national press. The subject was featured at least twice in the television programme 'Tonight'. In a number of places in Sussex there were displays of photographs of the affected landscapes, including some with pylons superimposed. The Action Committee wrote to the Ministers of Power, and Housing and Local Government, the former replying that he was very aware of feelings in West Sussex. Two deputations from the Action Committee attended committee meetings in the House of Commons; the Arts and Amenities Committee of the Parliamentary Labour Party, supplemented by representatives from the Party's Fuel and Power Group, in November; and with the corresponding Conservative Party representatives in January 1964.

The Action Committee's leaflet emphasized that they were fighting on a matter of principle and that the matter was of national, and not merely local importance. By arguing for the cabling of this westerly section the Action Committee had secured the support of organizations such as the Petersfield Society and the National Trust's Blackdown Committee, both of whom opposed the Blue route.

Rather surprisingly one major local preservation society, the Sussex Vigilants Association, after being represented at early meetings had decided during September 1963 to disassociate itself from the Action Committee. Although in complete sympathy with the aims of the Action Committee the Vigilants thought the emphasis on cabling mistaken. High cost ruled this out, so a more flexible attitude was needed if representations to the Minister were to have any chance of success. However, the Vigilants changed their attitude after the meeting of the British Association for the Advancement of Science, at Aberdeen in November, when high voltage electricity distribution systems not using overhead lines were discussed. The Vigilants interpreted this to mean that the pylons proposed for West Sussex might soon be obsolete, and wrote to the Prime Minister asking that the Sussex decision be delayed to allow these new ideas to be fully considered. A new if ill-founded confidence was aroused by this development, and in late November the Vigilants' President was speculating that further battle might be unnecessary as he had received a letter from the Minister of Power reporting that the Inspectors had been instructed to consider cabling possibilities. The protesting societies again presented a united front, although the co-operation between the Vigilants and the Downsmen appears throughout to have been somewhat subdued.

After brief mention in the national press in September 1963 the issue became almost dormant as a national news item until the Minister's decision was announced in mid-1964. An exception was an editorial in *The Times* in January 1964 entitled 'Pylons on the march', commenting favourably on the imminent designation by the Government of the Sussex Downs AONB, but noting the new threat to the area posed by the giant pylons planned to march across the Petworth and Petersfield area. Although pylons are not everywhere obtrusive, in the gentler West Sussex landscapes they would prove most objectionable, and 'the Minister should have another look at this threat to a specially designated area to see if some diversion or underground alleviation of part of the line cannot be designed. Otherwise why so designate an area?'[26] This prompted a letter from Horsham's MP, Frederick Gough, emphasizing the devastation which would be wrought by the proposal.

Before the Minister finally announced his decision the matter twice went before the Cabinet. Consultation between Power and Housing always took place on these cases, and as the Inspectors of the two Ministries were unable to submit a completely unanimous report, inter-departmental agreement would have had to be reached at some level. But there is no evidence that the two Ministers disagreed and the divergence in the report was of minor importance, capable of resolution at staff level. What was of concern were the political implications of the decision on the whole line, particularly of that part crossing the proposed AONB, given the active campaigning and influential spokesmen of the protestors. When the Cabinet showed an interest in the line, the Board was asked for further information, including the feasibility of an off-shore cable.

The Minister took an unusually long time considering the line; while the Inspectors' Report was dated October 1963, his decision was not published until 27 July 1964, when Parliament was in recess and the matter could not be raised there until after the October General Election. With the letter of decision, the Report of the Inspectors also became available to the public. The Inspectors recommended that the Minister should consent to the Red route in the Eastern Section, while in the Central Section he should refuse consent to both the Green and the Red and advise that the Board investigate the area to the north. In the Western Section they recommended that the Red route be withheld while the Board investigated the long Blue route, and at Buriton the Inspectors were themselves unable to agree on the exact place for the Downs crossing. The Minister decided in favour of the Red route throughout, rejecting the Inspectors' suggestions that alternatives be investigated on the grounds that, as the advantages claimed for these were too uncertain, it would be wrong to oblige the Board to investigate them and further delay this essential reinforcement of the system.

**Protesting the decision**

There would undoubtedly have been a considerable outcry against the decision, whichever of the routes was favoured by the Minister, if that decision provided for any overhead line in this area of Sussex. However, the rejection of the Inspectors' recommendations by the Minister added considerably to the general indignation aroused by the scheme, with both the Downsmen and Vigilants making the not uncommon allegation that the whole procedure was a farce. The timing of the decision gave rise to further criticism as no immediate parliamentary questioning was possible. It was reported that some landowners would certainly refuse wayleaves, the permission required from property owners for putting lines on and across their land. This would not greatly affect the ultimate outcome, but would introduce further delay for the CEGB.

As might be expected, there was much comment and criticism in the local press which was able to express great indignation about the rejection of the Inspectors' recommendations, thereby giving vent to local opposition without having to make invidious choices as between the routes favoured by various groups. With the amenity movement already alerted, the decision immediately transformed what had previously been essentially a source of local agitation into an outcry on a national scale, concerned not only with the Sussex Downs but also against pylons generally. Coupled with expressions of concern about the South Downs, there was also some discussion of the wider problems of the changing countryside. Perhaps more surprising was the concern shown in editorial comment of local newspapers in areas far removed from the South Downs, areas still bearing the scars of an earlier period of industrialization.

Meanwhile, people interested in the affected areas were preparing to make further protest. During September 1964 Midhurst RDC, disappointed that the Inspectors' preference for its Blue route was rejected, decided to write to the Minister of Power expressing dismay at the decision. Shortly before this, Petworth RDC had discussed whether to protest to the Minister for disregarding the advice of the Inspectors. Although strong feeling was evident the motion was defeated, perhaps because the Chairman made clear that the Minister alone was responsible for the decision. Generally, local authorities accepted that the battle was over and played little part in further protests.

By contrast the amenity organizations interested in the fifteen-mile Petworth to Petersfield route were not prepared to admit defeat. Following the Minister's decision the Action Committee had been revitalized as the Pylon Protest Committee to carry on the fight. Again the Society of Sussex Downsmen played the principal part, with their secretary, Mrs L. V. Ryan, being Secretary to the Committee. The threefold aims of the Committee were to inquire how the Minister had reached his decision, to secure full parliamentary debate of the whole scheme, and to seek a revision of the scheme to include underground cabling. Liaison between groups cannot have been strong during that August as, at the same time, the President of the Vigilants was expressing concern that no further courses of action had yet been suggested. Looking to the future, he considered a South of England Countryside Protection Committee was required to co-ordinate efforts to ensure proper consideration of landscape value. Such a united front would, he argued, be an effective counter to the Board's policy of divide and conquer (a reference to the line being dealt with in short sections) and would ensure that the total force of opposition was mustered. However, the Pylon Protest Committee was beginning to act, and in late August was advertising by letter to the local press the availability of an exhibition board for residents in the Midhurst area. This board, showing photographs of pylons, reports from the Committee, newspaper coverage and such like, was intended for exhibition at village fêtes and other local gatherings.

By early September the final campaign of the Pylon Protest Committee was getting under way. The Committee was asking supporters to canvass local prospective parliamentary candidates, and it was hoped that landowners would dispute wayleaves. A pamphlet issued a few weeks later by the Committee made it clear that the protests would be pursued in the next Parliament. The pamphlet, entitled 'Election manifesto of the pylon protest', was to be circulated to candidates in the forthcoming General Election. Whilst acknowledging that change and electricity must come to the countryside, it was argued that these could be achieved without destroying the beauty of this proposed AONB. Better planning was needed with determined efforts to reduce cabling costs. The pamphlet asserted that the matter would come up in the next Parliament and called on all election candidates to pledge themselves to support the Committee. The manifesto said that the Downs must be saved; 'They are a test case of the national will to preserve our heritage.'

Letters to the press were a favourite way throughout the dispute of bringing arguments before the public. The Chairman of the Harting Society, in a letter to *The Times* in late August 1964, called on the Minister to change his decision. But the Minister was not prepared to make any concession during the short remaining time of the Conservative Government. Despite the outcry over the fifteen miles, no change could be wrought and the Board maintained that, as pylons were so much cheaper, there was no alternative to an overhead route.

Following two articles to this effect by the Chairman of CEGB, published in *The Times* during September, Mrs Ryan replied on behalf of the Downsmen. She noted what the Chairman had said about the length of the consultations but held that the Minister was not justified in making urgency an excuse in rejecting his Inspectors' recommendation to investigate the alternative Blue route put forward at the Inquiry sixteen months earlier. This was only a small deviation to protect the Downs. She pointed out that it was equally important to preserve public confidence in the genuineness of public inquiries.

The Sussex Downsmen carried out a vigorous campaign in their own right as well as being the driving force behind the Pylon Protest Committee. Society membership provides a wealth of talented, articulate and influential people, and ready access to local MPs and other public figures is enjoyed. Lord Shawcross, the Downsmen's President, freely supported the Society and wrote to the new Minister of Power on its behalf. Early in September the Society approached the Council on Tribunals whose Secretary decided that there was a case for its consideration. However, the Council finally decided that the complaint about the refusal by the Minister to reopen the Inquiry was not legally valid, but that the case had raised certain points of principle relating generally to overhead line proposals and that these would be considered at a later date.

Meanwhile, in early October, a newsletter of the Sussex Downsmen promised further action and the establishment of a fighting fund for this and other work. In their campaign they were able to pursue their interests with a singleness of purpose impossible in the joint Protest Committee, particularly in pressing arguments favouring the Blue route. At the beginning of November another pamphlet was issued by the Downsmen as they started a 'winter offensive' aimed at enlisting the help of MPs. Over 10,000 copies of the leaflet were printed. They insisted that the line must be hidden from the Downs if it could not be placed underground, and called on objectors to write to the Prime Minister, Minister of Power and their MPs, insisting that the Blue route be adopted. The leaflet contained a series of questions relating mainly to the decision. Why was the careful, long and expensive Inquiry held if there was no time for implementing the findings? Why was the West Sussex Inquiry held nearly eighteen months after the neighbouring East Sussex Inquiry and after most of the route in Hampshire had been approved? Why did the Minister delay his decision for nine months and then say there was 'no time' and wait until the last week of Parliament to announce his decision? As it was then too late to get the matter debated, arrangements were being made to bring it to the attention of the new Government when Parliament reassembled, with a view to having the whole affair reopened. Mrs Ryan maintained that this was not a personal matter, she lived at Hove and most of the Society's 1,300 members did not live by the Downs, but they were anxious to ensure that they were not spoilt. At about the same time the President of the Vigilants was writing to Mr Fred Lee, the new Minister of Power. In an attempt to gain the sympathy of the new Labour Government for the anti-pylon campaign, he drew attention to the dismay caused by the previous Conservative Minister's decision and to reports of recent work that might make undergrounding more feasible. He asked for a halt to the pylon scheme while new methods of transmission were considered. A spokesman for the Ministry replied that once the Minister had given his decision it could not be revoked.

**Appeals to Parliament**

Discussion in Parliament occurred first in the House of Lords in November when, in his maiden speech, Lord Egremont (whose property, largely made over to the National Trust, was affected by the Red route) raised the matter of the contentious fifteen miles crossing the AONB. To preserve this fine countryside only a short section need be cabled, which would be a very small proportion of the Board's annual investment programme. Lord Egremont, formerly John Wyndam, Private Secretary to Mr Harold Macmillan, had already raised the matter once with the previous Minister of Power but had been 'met with a look like a stone wall with broken glass on top'. Unless people made a fuss there would be no alternative to overhead lines, and Lord Egremont declared he was 'making a fuss now; and if the Board's wayleave officer calls to see me at Petworth I shall have him thrown down the stairs',[27] a threat of violence which cannot have advanced the cause, and in the event he received two Board officials very hospitably. He concluded by asking the Government to cause this section to be placed underground and to increase research on cabling. Several peers supported the general case, although others, at least as much concerned with safeguarding the countryside, thought that the route chosen represented the best compromise possible. Replying for the Government, Lord Stonham drew attention to the mass of conflicting evidence and views, the CEGB's thoroughness in considering the possible routes, the cost of cabling and the amount of countryside scheduled for preservation. Time was only one of the factors mentioned in the decision letter for rejecting the Blue route.

In another maiden speech, this time in the Commons, the new MP for Horsham, Mr P. M. Hordern, made a similar plea for the Sussex countryside. He took the unusual step of voicing his opposition to the Board's plans during the second reading of the Finance Bill, and asked the Government both to halt the plans in the interests of the national economy, and also to encourage more research on underground cables. At the beginning of December, a number of Questions were put to the Minister asking him if he would review the recommendations of the Inspectors. Also, between November 1965 and the end of February 1966, there were several Questions about research into cables, the amount of cabling the Board was undertaking, and the public response to 400 kV lines already erected. In replying to a Question by Sir Hugh Lucas-Tooth about the possibility of making another government department responsible for amenity aspects of electricity schemes, the Prime Minister acknowledged that all this interest sprang from the Bolney-Lovedean decision.

Despite the Government replies in both the Lords and the Commons that the previous Minister's decision was unalterable, opposition in Sussex showed no signs of abating. Good coverage was still being given by the local press. One correspondent lamented the lack of activity in the Midhurst RDC area by the residents themselves, and hoped that the Sussex Downsmen could continue to protest against the undemocratic way in which the matter was handled by the Minister of Power. There had been no organized protest by the residents of Midhurst, and few from the whole area of the Midhurst Rural District which was to be designated as an AONB. The Midhurst and District Ratepayers Association had decided some months previously not to assist in any protest that might be organized, and the Midhurst Society probably felt it could not act because its area covered the Fernhurst Valley, through which the suggested Blue route would pass. In the event no organized protest had come from residents in the Midhurst area.

Early in December 1964 the Chairman of the Downsmen drew attention through the local press to a point in Ministry replies to the numerous recent written representations. The Minister had repeatedly stated that he could not grant any compulsory wayleaves without considering objections lodged by the landowner or occupier concerned, and had added that it would be possible for him to refuse wayleave consent. The Downsmen's Chairman said that it was in the interest of every landowner and occupier to make, in addition to an objection on his own behalf, a strong objection on the grounds that the route recommended by the Inspectors would be 'far less obtrusive and damaging to the countryside' and 'would affect fewer people'. A number of landowners followed this course of action, causing further delay to CEGB.

Undismayed by the rejection of representation made to the Ministry and in Parliament, the Downsmen continued their fight, and the January newsletter showed impatience with the Minister for refusing to 'rectify the gross blunder made by his predecessor'. The newsletter went to great lengths to show the widespread and responsible nature of its support: 'The whole outcry is to save the views of the Sussex Downs – not the fields or gardens of the residents.' Their members were drawn from all parts of the country, with many from London; comparatively few local voices had joined in the protest. A petition containing almost 300 names from Thornton Heath, a part of Croydon, had been sent to the MP for the Midhurst area. The Society claimed that many of its members were national figures, some of international fame, and it was anxious for the responsible government Ministers to realize this. Following the decision, the Downsmen had written to societies registered with the Civic Trust in order to sound out opinion amongst amenity groups, and all but three of 250 replies were in support of the protest.

However, the South Downs were not the only feature of great landscape beauty in the area, or the Downsmen the only organization concerned about the effect of pylons on such areas. The Blackdown Committee of the National Trust had watched with concern the correspondence and discussion and finally felt obliged to join the battle themselves. In a letter to the local press in January 1965 the Chairman of the Blackdown Committee drew attention to the importance of the Blackdown area with its fine panoramic views, and deplored the intrusion of pylons anywhere in the area. But, given the absolute necessity for the provision of electricity, the Red route was the least damaging, and a line in the Fernhurst Valley would mean that the Blackdown views were ruined in all directions. The concluding plea that cabling should still be most seriously considered illustrates why the earlier alliance with the Action Committee was possible, but that where the issue becomes mainly one of disputing alternative overhead routes unity tends to crumble.

As far as the overall planning permission for the Red route was concerned, the final rejection of the pleas for reconsideration came in the House of Commons on 22 January 1965. Mr Hordern spoke about the scheme in the adjournment debate, asking the Minister to revoke his predecessor's decision and order a new inquiry. He reported that he had seen almost 200 letters from civic societies throughout the country supporting the views expressed by the Society of Sussex Downsmen. The parliamentary Secretary to the Minister of Power replied that the correct procedure had been followed and consent given. Not only was the Minister unable to alter the decision, he did not wish to.

In the event, generous compensation terms were offered by the Board to induce landowners to sign voluntary wayleaves. However some property owners continued to resist and wayleave hearings were held during 1965 and early 1966. Despite his threats in Parliament, Lord Egremont signed a wayleave with the CEGB and negotiated a minor alteration within his estate, but his neighbour, Lord Cowdray, was among the ranks of those against whom the Board sought compulsory powers. As usually happens in cases where the planning consent has been strenuously contested and the issues already well aired, the Board obtained compulsory powers where necessary for the construction of the line on affected property, and the Bolney-Lovedean line came into operation late in 1967, eight years after route-planning began.

**In retrospect**

This overhead line dispute is unusual in the extent of interest generated both locally and nationally, and in the amount of effort and money individuals were prepared to expend opposing the various routes. With respect to private involvement it is almost a compendium of the features found, variously, in the sixty-odd CEGB line inquiries since 1958, but the intensity of involvement makes it distinctly unrepresentative. None of the routes proposed (official or otherwise) was free from private objection while some, minimizing this type of concern, tended to increase broader-based amenity objections. Whilst planning authorities might hope to obtain short underground sections for outstanding natural features, no such amelioration could be hoped for when fifteen miles of line was involved, even though the countryside was of great merit. Given this limitation, the County Council and National Parks Commission views coincided in all but the Central Section. For the major public authorities no great principle was at issue, and the Inquiry is memorable rather for the labyrinthine wrangles among private individuals, amenity societies and district authorities.

Much offence was given by the Minister's rejection of the Inspectors' recommendations. By consenting to the Board's route throughout did the Minister merely take the easy way out, as was alleged at the time? Without embarking upon the hazardous task of evaluating the merits of the various tracts of countryside,[28] some explanation can be advanced by examining the participants, their interests, and the particular objections they made, in relation to the Inspectors' findings. The Eastern Section presents few problems as, apart from Horsham RDC and one group of private objectors, the weight of evidence at the Inquiry favoured the Board's route. With the National Parks Commission, local amenity bodies, all but one of the Parish Councils, many private objectors – and by implication the County Council – all supporting the Red route, the Minister had no difficulty in accepting his Inspectors' recommendations. An additional, if minor, advantage was that the allegations concerning Horsham RDC could be more easily forgotten.

The Minister's decision in the Central and Western Sections raises more difficulties, for of the seven inquiries since 1958 in which a Minister has over-ruled the Inspectors this is the only case where it has been to the advantage of the Board rather than to objectors. In the Central Section the Inspectors suggested that a wholly new route be investigated, rejecting both the Red and Green routes for which formal application had been made. The Minister was impressed by the care taken by Miss Crowe in selecting the Red route, was conscious of the long delay investigation of a new route would entail, and was aware that no one at the Inquiry had canvassed such a route to the north. As between the Red and Green routes, he accepted the views of the Board's consultant, the NPC and the Nature Conservancy rather than those of the local authorities. The irony is that if there had been less disagreement between the local authorities the Board might not have appointed Miss Crowe and been content to apply only for the Green route in the Central Section.

Delay was again involved in the acceptance of the Blue route recommended by the Inspectors in the Western Section. The acknowledgment by the County Council that the Red route was the best available would almost certainly have influenced the Minister, as did the fact that, earlier, expensive cabling of a 132 kV line had been required to preserve the Fernhurst Valley. In rejecting the Inspectors' recommendation, the Minister preferred the arguments of the Board, the NPC and the CPRE and some local amenity groups to those of the RDC, the Sussex Downsmen and various other local amenity bodies. Besides offending supporters of the Red route, the large number of private objections received from the Fernhurst Valley to the unofficial Blue route indicated that a formal application would be strenuously opposed, thus starting the whole process again. The alternatives favoured by the Inspectors both here and in the Central Section were longer than the Red route, another factor relevant when respective advantages are finely balanced.

Much was made by the Sussex Downsmen and others of the Minister's reference to the further delay any formal consideration of the Blue alternatives would entail. Although the Inspectors recommended investigation, the Minister was correct in arguing that the full weight of objections against the Blue routes had not been voiced at the Inquiry. Provided that cabling the fifteen miles was still ruled out, it is not easy to avoid the conclusion that even after a second inquiry the balance of advantages would still have lain with the Red route. An additional pressure for a definite decision by the Government at that stage was, of course, the continuing preoccupation with economic development, the South Coast line being a corollary of that policy. However, a decision announced too late for parliamentary debate before dissolution suggests a lack of courage. Yet the decision itself was courageous, bearing in mind the political and personal links between the Conservative Party and this area.

Private individual action on the scale evident in this dispute is somewhat self-defeating. Both the Board and the Minister are sympathetic to private owners seeking alleviation, but will not acquiesce in proposals that merely transfer the line to another person's property. The difficulty is that counter-proposals or opposition coming from some people must be answered by those to be affected or their case may go by default. Where the case cannot be buttressed by reference to wider values, such as the property being of historic value, gardens that are open to the public, or the amenity value to the generality, the best that can often be gained is monetary compensation. There is a danger that private protest on this scale may even damage the cause of more general countryside conservation, as amenity protests may be linked in the popular mind with preservation of middle-class privilege. One cannot but sympathize with the fate, increasingly common for those who have bought or built property with high amenity situations, but the considerable amount of energy spent in trying to safeguard one set of property owners at the expense of another contributed little to the wider public interest in this case.[29]

By their nature amenity organizations are speaking for a wider interest. The President of the Vigilants, commenting on impressions he had gained while attending protest meetings before the Inquiry, went out of his way to emphasize that the Association would never defend private interests. He attacked those who join amenity societies to further their own personal interests rather than those of the area as a whole. For the smaller society interested in preserving an immediate locality the problem must be to decide when it is pushing local interests at the expense of wider amenity values. Thus greater attention was given by the West Sussex County Council to a larger group such as the Sussex Downsmen than to the more local preservation societies. The key to the failure of the Downsmen in the Western Section arose from the particularity of their interests rather than from any deficiencies in their campaigning.

The admission at the Inquiry that the Society was a pressure group for the South Downs provides the explanation; there was not sufficient unanimity about the absolute priority of the Downs to persuade the Minister to change the existing policy of protecting the Fernhurst Valley from overhead power lines.

Events in the west of the county show a continuing problem for amenity groups protesting against unavoidable developments, and demonstrate the need to construct stable coalitions so as to present a united front against the proposals. The Action Committee of 1963-4 maintained its unity by pressing for underground cabling, a solution equally acceptable to the opponents of the Blue and Red routes. After the decision was announced, publication of the Inspectors' recommendations provided powerful support for the Sussex Downsmen in their fight to keep the pylons away from the Downs. Handicapped by the need to maintain an alliance of geographically disparate associations, the Pylon Protest Committee, like its predecessor, still sought cabling. With the well-organized Downsmen being the driving force, the Protest Committee became less important as the Downsmen put increasing effort behind any solution able to protect their basic concerns. Eventually this brought open disagreement between them and the Blackdown Committee of the National Trust. The Pylon Protest Committee did serve to alert and involve people in the dispute and in fact endured beyond this crisis to form the basis of a collective organization to deal with similar problems. Arising out of this dispute the amenity groups in Sussex have now organized on a federal basis with a joint committee to deal with county-wide issues. While concerned with all kinds of issues, electricity projects still cause anxiety and the groups concerned played a major part in frustrating CEGB's original plan for a line between Chichester and Lovedean in the late 1960s.

Another lesson learnt is that it was a grave tactical error on the part of the amenity societies to mount the main public campaign after the Inquiry. Partly this arose out of the then fragmented structure of the amenity movement in Sussex. The time to influence a decision is before it is taken, preferably at the stage when the major authorities are still working out their own attitudes – often before the proposed development has been formally announced.

## Notes

1 For a discussion on the procedures and issues involved in the granting of consents for overhead electricity lines see H. R. Burroughes, 'Political and administrative problems of development planning: the case of the CEGB and the supergrid', *Public Administration*, vol. 49, Summer 1971.
2 400 kV double circuit heavy duty lines are carried on steel lattice towers about 165 feet high, with twenty-four conductors suspended in groups of four wires from three pairs of side arms, plus an earth wire. Horsham's MP described them as gaunt and ugly monsters bestraddling the area. Stronger emotions are occasionally aroused; a Barnet UDC official once described them as 'these three-headed skeletons with their trailing entrails'.
3 A Public Inquiry is mandatory where a planning authority objects, but discretionary in the case of other classes of objectors. See also note 23.
4 *Sunday Times*, 6 September 1964.
5 Inspectors' Report, October 1963, p. 40; see also Public Inquiry, day fifteen, p. 12.
6 Letter quoted in *West Sussex County Times*, 8 March 1963.
7 Public Inquiry, day fifteen, evidence of Mr H. C. N. Henry.
8 *West Sussex County Times*, 22 February 1963.
9 *West Sussex Gazette*, 7 February 1963.
10 *West Sussex County Times*, 22 February 1963.
11 Ibid., 8 March 1963.
12 Inspectors' Report, October 1963, p. 32.
13 Public Inquiry, day fifteen, p. 26.
14 Ibid., 3 May 1963.
15 Public Inquiry, day fifteen, evidence of Mr H. C. N. Henry.
16 Ibid.
17 Quoted in *West Sussex County Times*, 3 May 1963.
18 Inspectors' Report, October 1963, p. 37.
19 *West Sussex Gazette*, 23 April 1963.
20 Public Inquiry, day fourteen, evidence of Mr S. W. Fowler.
21 *West Sussex County Times*, 22 February 1963.
22 For an examination of these inquiries see H. R. Burroughes, 'Public inquiries and large scale development', *Public Law*, Autumn 1970.
23 The nature and role of public inquiries are exhaustively discussed in R. E. Wraith and G. B. Lamb, *Public Inquiries as an Instrument of Government*, Allen & Unwin, 1971.
24 The Inquiry evidence is summarized in the Report of the Inspectors to the Minister of Power, October 1963.
25 These were: Blackdown Branch of the NT, Buriton Parish Council, Durford Wood Trust, Harting Preservation Society, Lodsworth Preservation Society, Midhurst Preservation Society, Midhurst RDC, Petersfield Society, Society of Sussex Downsmen and the Sussex Rural Community Council.
26 *The Times*, 18 January 1964.
27 House of Lords Debate, vol. 261, cols 547-68.
28 See R. Gregory, *The Price of Amenity*, Macmillan, 1971, ch. 1, for a general analysis of issues involved in this type of amenity dispute.
29 See T. Aldous, *Battle for the Environment*, Fontana, 1972, pp. 238-40.

# Chapter 5

# The politics of Manchester's water supply 1961–7

Susan J. Dolbey

> It's becoming almost a game. You get a map of the Lake District, preferably one showing a bit of coastline and some contours, and then you work out how Manchester can solve its water problems without ruining the place. It doesn't matter whether you're an engineer for nobody's going to quibble about a few million gallons – or a few million pounds for that matter.
>
> A. H. Griffin, *Lancashire Evening Post*, 19 September 1964

This chapter is a case-study of Manchester's attempts to obtain additional water from the Lake District in the period 1961–7.[1] This controversy has been regarded as one of the more interesting disputes involving the desire to preserve the environment and at the same time to accommodate the basic needs of a highly populated country. The outstanding beauty of the Lake District National Park is an accepted fact, as is its abundant rainfall. The conjunction of these two factors provides the source of this passionate conflict about the use of the area's natural resources.

### Summary of events 1961–7

In the autumn of 1961 Manchester decided to promote a Private Bill to abstract water from Ullswater in the Lake District and to create a reservoir in Bannisdale. The Corporation stated that its current supplies from Longdendale in the Peak District and Thirlmere and Haweswater in the Lake District would no longer be adequate. Manchester was opposed by local authorities in the Lake District, statutory bodies, a great number of national and local amenity organizations, local landowners and individuals from all over the country. The Bill was debated in the House of Lords and defeated on second reading.

In an attempt to reconcile the differences between the Lake District and Manchester, the Government convened a conference of interested local authorities and statutory bodies. The Jellicoe Conference, as it was called, met for a year (March 1962 to February 1963) and established the need and possible sources for additional water supplies which were suitable from an engineering and amenity point of view.

In the spring of 1964 Manchester announced its intentions of seeking a draft Order for permission to abstract from Ullswater and Windermere and to duplicate the aqueduct in Longsleddale. The Corporation was opposed by many of its 1961-2 opponents and a number of new local preservation societies.

Following a public inquiry, the Minister decided in 1966 that Manchester would be permitted to abstract from Ullswater and Windermere, with safeguards, and would not be allowed to duplicate the Longsleddale aqueduct.

## The first period 1961-2

In this period political rather than technical skills were important. Significantly, they were important because the opponents realized that their strength was in political action not technical argument and arranged that the battle would be fought on their grounds. Equally significant is the fact that Manchester did not make the most of its technical potential and was not politically astute enough to ensure that the dispute would be settled on grounds favourable to the Corporation.

### *Manchester*

With hindsight it is easy to see that in the first period Manchester underestimated the opposition there would be to its proposals and underestimated the political skills of its opponents. It seems likely that the Corporation's ultimate mistake was to allow its Bill to be heard first in the House of Lords.

However, early in 1961 Manchester had reason to be optimistic. The Corporation had had powers since 1919 to increase the size of the aqueducts in the Haweswater system, including larger pipes from Heltondale which was only six miles from Ullswater. Bannisdale had been suggested as the most suitable source for Manchester in the Vail Report published by the Ministry of Health in 1949.[2] The Corporation had chosen the obvious sources. Both Ullswater and Bannisdale were close to existing aqueducts; they fitted into the Corporation's existing water system; they were cheap and they could be completed quickly. The Corporation needed to act quickly. The demand for water was accelerating and this, coupled with the fact that Manchester's water-supply area had been enlarged over the past few years, meant that Manchester was consuming water at a far greater rate than had been anticipated.

The indication of government support and the unanticipated need for speed partly explains why the Corporation prepared its actual plans in such haste. The consulting engineers were not hired until July 1961. The decision to promote a Private Bill requesting permission to extract water from Ullswater and create a reservoir in Bannisdale was formally approved by City Council on 1 November 1961. Technical studies of the dam site in Bannisdale were still being done in January 1962.

Initially the Corporation was so confident that little thought seems to have been given to ensuring that the proposals were carefully announced and promoted. Manchester regarded some opposition from amenity organizations as inevitable, troublesome, and bound to be ineffectual. It was the Corporation's view that as any scheme would be opposed by such groups as the Friends of the Lake District there was no point in trying to accommodate them.

Manchester was totally unprepared for the massive outcry that came not only from these amenity groups but also from virtually every statutory and local authority in the Lake District and from individuals all over the country.

Manchester did much to increase the volume and size of the opposition by mishandling the presentation of its plans. The first indication many in the Lake District had of Manchester's plans was a very vague press release:[3]

> Among the proposals now being investigated are dams and reservoirs in the Bannisdale and Borrow Beck areas, north and east of Kendal; a possible intake scheme from Ullswater and possible catchments north and west of the existing Lake District sources.

The ambiguity of this announcement frightened many; the Corporation further antagonized many bodies because this announcement was sent only to those Lake District bodies who had opposed a previous project of the Corporation. As a result all statutory and local authorities in the Lake District were left to find the announcement in the press.

The Corporation was seen as an autocrat disregarding local feelings and taking what it wanted. Manchester failed to explain its proposals clearly. For example it was widely believed that Manchester intended to turn Ullswater into a reservoir like Thirlmere and Haweswater.

By the time the Corporation did consult with Lake District

bodies, public and official feeling was against Manchester almost regardless of what the Corporation proposed to do. Little was accomplished in these consultations and they were sometimes characterized by a good deal of acrimony. They failed for a number of reasons. The Lake District authorities were afraid of prejudicing their case; Manchester's schemes were never sufficiently prepared for others to judge the proposals and suggest amendments of either a technical or amenity nature; the Corporation failed to explain convincingly the difference between a dam and a weir; and the opponents were not interested in bargaining for an increased water supply, as they placed a higher value on the preservation of the area.

Manchester's greatest political mistake may have been its acquiescence to the Bill being read first in the House of Lords. Although the Corporation was informed in advance, it raised no objection. It seems likely that had it been read first in the House of Commons, the Government would have supported the Bill.

Manchester's tactics for the debate were essentially negative. It had few friends in the House of Lords and therefore could have little hope of getting large numbers of 'backwoodsmen' to support its Bill. Instead, it attempted to dissuade Lord Birkett from moving a motion to defeat the Bill at second reading and attempted to persuade the Government to use the Whips. Although the Government refused to put the Whips on in the House of Lords, it did attempt to persuade Lord Birkett to withdraw his motion. In the event, neither Manchester nor the Government changed Lord Birkett's mind.

*The opponents*

In this period, 1961-2, the opponents quite rapidly realized their strengths and weaknesses and succeeded in making the most of their assets.

Almost all the affected local authorities, including the County Councils and the Lake District Planning Board, opposed the scheme. The Ullswater Preservation Society was revived by a group of local landowners, and various established amenity organizations such as the Council for the Preservation of Rural England (CPRE), the National Trust and the Friends of the Lake District took up the fight.

The opponents sought to advance their cause in a number of ways. Initially there had been the expectation that an engineer, like counsel, would interpret and present the facts in the manner desired by the client. They were soon to find that the water engineering consultants used objective professional standards to judge the merits of Manchester's proposals. Their bias was towards good engineering schemes at an economical cost. The opponents' consultants and Manchester's engineers viewed the situation in the same light. The water engineering consultants engaged by the Westmorland CC, the Cumberland CC, the National Parks Commission and the Lake District Planning Board could not fault Manchester's proposals on engineering grounds; tentative proposals for alternative sources of water were not acceptable to the Lake District bodies; and the consultants could not formulate alternative schemes as Manchester had not allowed sufficient time. Before the debate the opponents realized they could not depend on the evidence of consulting engineers to defeat the Bill.

In this period the lack of engineering support was to be an asset to the opponents. Their energy was not divided between technical and political opposition. More importantly, as no alternative sources were suggested, no part of the opposition was threatening another part and the opposition remained united against Manchester. Even though there was no strong feeling about Bannisdale, the opponents opposed the Manchester proposals *in toto* as they wished to discourage Manchester from ever returning to the Lake District. The general Lake District reaction was that Manchester must be stopped from taking any further water from the Lake District National Park. It was feared that these new works would destroy the area's amenities and would reduce access. It was felt by many that this scheme would be the first stage of a much larger water project. Local landowners were dismayed by the threat to their property.

Thus the opponents were united by purpose. Co-ordination was also greatly facilitated by the overlapping membership of the various bodies and by friendship networks.

Although the opponents were weak on technical support, they had considerable success in amassing and showing public support. This was achieved by enlisting the help of other organizations, holding public meetings, obtaining press coverage, writing letters to editors of local and national newspapers and circulating a public petition. At a public meeting those present 'loudly applauded descriptions of Manchester Corporation as contemptuous, greedy, monstrous, idiotic, imprudent, and autocratic'.[4] The Lake District papers gave much coverage to the issue, wrote sympathetic editorials, and published many letters to the editor. The national press also covered the controversy and published letters from prominent persons. The Ullswater Preservation Society circulated a petition which collected over 500,000 signatures.

An attempt was made by the Manchester Centre of the National Trust and the Ramblers' Association in Manchester to defeat the

Bill at the statutory town meeting. Had they succeeded the Council would have had to decide whether to poll all the ratepayers in the hope of reversing the decision. However, this attempt was begun too late to judge whether this was a useful technique to use. Only six of those present at the town meeting opposed the Bannisdale scheme; the Ullswater scheme was approved by a three-to-one show of hands.

Through letters, delegations, and through the Lake District MPs, the opponents attempted to influence both the Minister of Housing and Local Government and his civil servants. However, in general, the Government did support this Bill although it refused to put the Whips on in the House of Lords.

The opponents' success in this first stage was largely due to careful organizing of the House of Lords debate. One of the first actions of the interested Lords and local MPs was to persuade Lord Birkett[5] to open the debate. During January 1962, Lord Lonsdale[6] held organizing meetings with those peers who intended to participate in the debate. These peers feared that if the Bill went to committee they would lose. Therefore they thought it was essential that the issue should be debated in principle in the whole House. Hence it was important to get as many sympathetic peers to the debate as possible. To encourage attendance many opposing bodies sent circular letters presenting their case to Lords and urging them to attend. Lord Lonsdale also supplied the opponents with lists of active peers and generally advised them about which peers to solicit. He suggested that the National Trust should write to at least their obvious friends, as well as writing to the members of the Government. It appears that the Cumberland CC, the Westmorland CC, the Lake District Planning Board, and the Ullswater Preservation Society circularized those on Lord Lonsdale's list. The National Trust wrote to its 120 members in the House of Lords. Mr P. J. Liddell, Chairman of the River Eden and District Fisheries Association, circularized fifty peers – those in the Salmon and Trout Association and others with whom Mr Liddell fished. The River Boards Association wrote to fourteen Lords. Mr Stafford Howard of the Ullswater Preservation Society wrote to old army friends and to his fourteen cousins in the House of Lords. The CPRE sent a short circular to members of the Independent Association of Unionist Peers. The Country Landowners' Association's memorandum was sent to about 300 peers – its members and the members of the Independent Association of Unionist Peers.

In all there were seventeen formal petitions against the Bill. These represented objections from twenty-seven organizations and four private individual petitioners.

**The debate**

The debate was very interesting if illogical. The essence of the debate was whether the Bill would be sent to a Select Committee which would hear and judge the arguments for and against Manchester's proposals. Accordingly, both sides presented arguments for and against the Select Committee dealing with the proposals. In addition, arguments of the sort that would be heard by a Select Committee were presented and a number of emotional appeals were made.

The Chairman of Committees explained that to give second reading to a Private Bill was not to give approval in principle – the Select Committee would consider both policy and details.

Lord Birkett however maintained that there were some Private Bills 'where the point of principle is so grave and so great that this House ought not to derogate from its power and say "Let it go to a committee".'[7]

This, he claimed, was one such Bill but he did not clearly state the principle. Rather he appeared to define three principles:[8]

> Shall the Manchester Corporation be permitted to invade Lakeland for the third time, to impound its waters, to pour them into its aqueducts, or not? That is the question of principle about which your Lordships are asked to make up your minds.

> The great overriding principle which never operated before in the case of Thirlmere or in the case of Haweswater arises now – namely, are we going to allow this in a National Park?[9]

> Produce the hydrological data on which the House can come to a proper decision. Until that is done, you have no right whatever to invade the sanctity of a National Park. The principle will be invaluable if it is established by the House.[10]

Those in favour of second reading argued that the only just action was to judge the case on its merits. Lord Hailsham made this point:[11]

> I have a rooted objection to trying cases before you have heard the evidence; and until today I had supposed that the noble and learned Lord on the cross benches [Lord Birkett] shared my prejudice.

Lord Birkett's reply was that this was a matter of principle. He was then taunted by Lord Alexander of Hillsborough, 'What is he really afraid of in sending this to be examined with full evidence by a Committee?'[12] This indeed was the issue.

Lord Birkett won. By a vote of seventy to twenty-six the Lords denied a second reading to the water section of Manchester's Bill. Many factors could have influenced the outcome of the debate. The most important, however, is clear. The opponents succeeded in persuading many 'backwoodsmen' to attend the debate and vote against Manchester's proposals.

In seven divisions preceding Manchester's, and in the seven following, the average number of voters was eighty-six. In this division 106 peers voted.

As the number exceeded that usually voting in divisions it is important to discover the number of Lords who made this their only debate of the fifteen ('backwoodsmen') and whether the remaining Lords were regular or casual voters. The largest group of voters were 'backwoodsmen'. Of the twenty-nine 'backwoodsmen', twenty-five opposed second reading; four favoured it. Had the vote been among only those Lords who voted in ten or more of the fifteen divisions Manchester would have won. Clearly 'backwoodsmen' were decisive in this debate as they voted disproportionally in favour of refusing second reading.

It was widely rumoured in the press, and the Earl of Crawford and Balcarres stated in the House of Lords, that 120 members of the House of Lords were members of the National Trust. Indeed Manchester thought that some of the peers might have been unduly affected by statements of the National Trust. However, as we have seen, it was not only the National Trust that wrote to peers urging them to attend. Those responding to appeals from the National Trust and the Cumberland and Westmorland CCs accounted for 55 per cent of all those in favour of defeating Manchester's proposals; and for 64 per cent of all 'backwoodsmen' who voted against them. National Trust members accounted for 52 per cent of these 'backwoodsmen'. Thus, although it cannot be proved that these appeals alone brought most of Lord Birkett's support, they probably had some effect especially in bringing out 'backwoodsmen'. In contrast, of the twelve peers solicited by Manchester five (14 per cent of those that voted against Lord Birkett's motion) voted in Manchester's favour.

Table 5.1 gives various factors about those peers who voted in this debate.

The basic conclusion is that Manchester's proposals were defeated by 'backwoodsmen', 'backwoodsmen' whom the opponents had persuaded to attend. The Lake District was delighted. By mobilizing their considerable political forces in the House of Lords they had won the game. Little thought had been given to long-term strategy or to the result of defeating the Bill. Because the opponents had little understanding of Manchester's need for water or of the

TABLE 5.1 Information about voters

| For Lord Birkett's motion | Against Lord Birkett's motion |
|---|---|
| 39 Conservatives | 23 Conservatives |
| 8 Labour | 9 Labour |
| 11 Liberal | 9 Liberal |
| 12 cross bench and unknown | 2 cross bench and unknown |
| 21 peers of creation (58%) | 15 peers of creation (42%) |
| 26 post-1900 peers of succession (76%) | 8 post-1900 peers of succession (24%) |
| 23 pre-1900 peers of succession (64%) | 13 pre-1900 peers of succession (36%) |
| 25 backwoodsmen | 4 backwoodsmen |
| 25 members of the National Trust | 5 members of the National Trust |
| 7 members, at least, of other opposing organizations | Chairman of British Waterworks Company |
| | Chancellor of Manchester University |

technical problems of producing feasible schemes, they rather vaguely hoped that Manchester would either give up its attempts to obtain more water or would get its water from unspecified sources far from the Lake District.

## The Jellicoe Conference

These hopes were short lived. In late February 1962, the Ministry of Housing and Local Government convened a conference of Manchester, some of the authorities Manchester supplied with water, Lake District authorities, and the National Parks Commission. In broad terms the Conference was to assess Manchester's need for water, to suggest sources for water, and to comment on these sources from both an amenity and technical point of view.

The Conference lasted a year. It showed to all that the Corporation had had a good case. It needed water quickly; few schemes could be completed in time; those that could be were in the Lake District; and the sources that could be most quickly developed were Ullswater and Windermere. The Conference found the Windermere scheme acceptable on amenity grounds but had great difficulty agreeing on the effects that the Ullswater scheme would have on the environment.

The Nature Conservancy was asked to comment on Ullswater. Unfortunately for the Conference, the Conservancy's report was vague and uncertain and gave support and opposition to both sides. Many summaries of the Conservancy's views were drafted for the Amenity and Planning Committee of the Conference before the following version appeared in the final report of the Jellicoe Conference:[13]

> The majority conclude that the report corroborates their view that it is possible that vegetation around the lake-side might suffer and this could cause some injury to amenity. Some members, however, did not agree with this conclusion and point out that the Conservancy's report itself suggests that important permanent changes are improbable.

The Lake District had a majority of members on the Committee; Manchester's members were the 'some members'.

The Conference was reasonably successful in achieving its aims. It clearly stated the facts that were acceptable to both sides; it investigated all the obvious sources; and it eliminated certain areas, such as Bannisdale, on amenity considerations. However, the Conference missed some of the less obvious schemes (those for the River Leven) and it failed to reach agreement on the one source which was the crux of the matter – Ullswater.

## The second period 1963-7

Many of the actors and the arguments were similar in the second round of Manchester's attempts to obtain water from the Lake District. However, the differences were of more importance than the similarities. The distinctions between the two periods arose from the differences in Manchester's plans, from the lessons the participants learned from the 1961-2 disputes, from the facts revealed in the Jellicoe Report, and later from the information in a report written by Manchester's consulting engineers. In this second struggle more opposition groups were established, and the emphasis of the arguments changed. The greatest difference was that much more attention was paid to the technical aspects of water engineering by both Manchester and the opponents.

### *Manchester*

Manchester had learned a lot from the first dispute. This time the city set the tone of the dispute and it chose a method of obtaining approval where the emphasis was on technical support rather than political connections. It also managed to delay the process quite considerably. It will be argued in the conclusions that a key factor to such a dispute is timing.

During 1963 Manchester hired three firms of consulting water engineers to prepare new schemes. It also decided to secure the necessary powers to implement its schemes by an order made under the Water Act 1945 rather than by a Private Bill.

The consultants submitted their report in February 1964.[14] In essence they recommended two short-term schemes – abstraction of 25 mgd (million gallons a day) from Ullswater and 20 mgd from Windermere. The report also suggested that the best long-term scheme was to abstract water from Windermere and store it in the Winster Valley.

Subsequently the Corporation, on 25 May 1964, agreed to promote a Water Order for 45 mgd from Ullswater and Windermere and an aqueduct from Haweswater to Watchgate. (This became known as the Longsleddale aqueduct.)

Manchester's major and most successful tactic in the struggle for more water was to produce schemes that were good technically and were also relatively free from amenity objections. Although the Corporation still sought to abstract water from Ullswater, the new plan eliminated the weir that had produced much of the criticism in 1961-2. Manchester's other scheme, Windermere, had been, in essence, approved by the Jellicoe Conference. Both schemes had been carefully considered: three of the country's best water engineering consultancies spent a year as part of the Technical Committee of the Jellicoe Conference, and then another year on their own preparing them. It was also expected that the discussion of alternatives in the consultants' report would prevent the Lake District from accusing Manchester of not looking for them. The Corporation could hope for victory on the merits of the case.

Manchester did relatively little political lobbying in the period leading up to the Public Inquiry. This is not surprising as it was to be the Minister's decision on the draft Order that was important. In his quasi-judicial position the Minister is not expected to meet with any of the interested bodies before he makes his decision. The Ministry had no discussion with the Corporation about what sources should be considered and it made no official inspections before the Public Inquiry. Manchester did explain its proposals to local MPs. Newspaper reports of the meetings varied, but it appears that Labour Members generally favoured the plans, whereas the Conservatives were more cautious.

From Manchester's point of view, the consultations in 1963-5 were better handled than those in 1961-2. In fewer cases was lack of consultation a reason for opposing Manchester. This time a number of safeguards and agreements were reached between Manchester and interested organizations, particularly the River Boards.

Manchester learned from its past mistakes not only on the subject of consultations but also in the field of public relations. At the beginning of 1964 the Waterworks Committee hired a firm of public relations consultants. The consultants held press conferences, advised the Corporation on public pronouncements, and placed advertisements in the press. But some of the former difficulties continued. At the first press conference organized by the public relations firm, one of Manchester's water engineering consultants stated that he was a little tired of amateurs. He said that people who understood the least about water engineering had the widest public say when it came to controversy. This of course annoyed many Lake District enthusiasts and prompted the following letter to *The Times*.[15]

> You may be sure, Sir, that of the half million or so of us in the Lake District many are heartily tired of the jackbooted arrogance of Manchester's professionals.

To counteract this kind of image and many fears of the Corporation's schemes, paid bulletins were placed in the Lake District press.

The nine bulletins emphasized different points, but together told the reader that Manchester would not pump from Ullswater when the lake was low; that there would be no more reservoirs, or interference with beauty, or restrictions to access; that temporary inconvenience would be minimal; that the Corporation made no profit from water; and that Manchester did not intend to flood the Winster Valley. On the positive side, the bulletins maintained that the need was urgent, that Manchester supplied a large area with water, that this was 'a first class water conservation scheme', that pumphouses and pipelines would be underground and silent, and that the new treatment works would permit public access to Haweswater.

It is clear that in 1963-5 Manchester really tried to win. The Corporation made a real effort to consult and to improve the public's image of Manchester and its plans. However, of all its weapons the most powerful was the excellence of the proposed schemes. The next section will show that the merits of the undertaking's plans were the biggest stumbling block of Manchester's opponents.

### *The opponents*

The opposition in 1963-5 differed from that in 1961-2. Those that opposed in the first period did so again and many of the same arguments were advanced. However, in this period there were also new opponents and new arguments. The opponents had difficulty in presenting a united front and they had difficulty in fighting what was to be essentially a technical battle.

Four new groups were formed during the 1963-5 period to oppose proposals specifically connected with the Corporation's search for new sources: the Morecambe Bay and Kent Estuary Preservation Society, the Winster Valley Preservation Society, the

Lakes Defence Co-ordinating Committee and the Windermere and Leven Valley Preservation Society.

The Morecambe Bay and Kent Estuary Preservation Society was formed in response to a remark in the Jellicoe Report that storage could be created in the estuaries of the Kent and Leven Rivers. The Society was pleased that Manchester's consulting engineers' report rejected the Morecambe Bay barrage. However, the Society stayed in existence and continued to fight Manchester because: 'It is obvious that Winster's fight is our fight . . . because once Manchester gets a foothold there the Leven and Kent Estuaries will, in years to come, be put forward as the logical extension of the scheme.'[16]

The Winster Valley Preservation Society was formed in response to the long-term scheme suggested in the consultants' report for the creation of a reservoir in the Winster Valley.

The Lakes Defence Co-ordinating Committee was formed following a meeting in June 1964 called by the Ullswater Preservation Society. All associations that might have had even the slightest interest in opposing Manchester's plans were invited to discuss the best way of objecting at the Public Inquiry, and how to support one another without duplication of effort. The Co-ordination Committee was originally composed of the Ullswater Preservation Society, the Morecambe Bay and Kent Estuary Society and the Winster Valley Preservation Society.

Shortly after the Co-ordinating Committee was established, the Windermere and Leven Preservation Society was formed and became the fourth member of the Lakes Defence Co-ordinating Committee. The Windermere Society was formed partly to defend Windermere against Manchester's proposals, but primarily because members of the other local societies were advocating abstraction from Windermere.

In this second period Manchester's opponents had different attitudes. The River Boards felt that as long as a scheme did not increase pollution, or damage drainage or fisheries, they had no grounds for objection. The Lancashire River Board had no objection to Manchester abstracting from Windermere provided there were safeguards. With the same proviso, the Cumberland River Board did not oppose abstraction from Ullswater. However, it strongly opposed the popular alternative of taking water from the River Eamont.

The Lancashire CC, with a foot in both camps, did not oppose except to obtain safeguards. Indeed the county publicly supported Manchester's plans for not only Windermere but also Ullswater. Both Cumberland and Westmorland County Councils saw Manchester's proposals as threats to the counties – to the amenities, to the roads and to the future. Although Westmorland was willing to consider Windermere if there was no better alternative both councils were completely opposed to the Ullswater scheme, although both were willing to negotiate to obtain safeguards.

Although both the Lake District Planning Board and the National Parks Commission felt more strongly about the Ullswater proposal, they decided to oppose Manchester's plans for both Ullswater and Windermere.

The national amenity societies, including the Friends of the Lake District, opposed Manchester's schemes 'on principle', opposed all alternative schemes in the Lake District and all favoured an alternative scheme for abstraction from the River Leven at Greenodd.

The local preservation societies took a more parochial view. Quite naturally, they tended to oppose the schemes which affected their local areas as a matter of priority, although the Lancashire CPRE took a curiously inconsistent stance. On the one hand, like the Lancashire CC, the branch took account of the fact that many Lancashire residents were dependent upon Manchester water; it also concluded that Manchester's plans for Ullswater and Windermere would not be harmful. On the other hand, it decided that to put another aqueduct through Longsleddale could not be tolerated.

The main role of the Lakes Defence Co-ordinating Committee was to keep peace among its members. Because of this it made few positive or strong statements.

The Windermere and Leven Preservation Society and the Ullswater Preservation Society shared common views: no lake abstraction, fear for the future, objection 'on principle', and objection through political channels. The Morecambe Bay and Kent Estuary Preservation Society and the Winster Valley Preservation Society also shared beliefs: lake abstraction would be permissible, no reservoirs or barrages would be allowed, objection through technical comments about Manchester's plans, and objection by suggesting alternative schemes that were technically feasible.

The attitudes of all opponents were influenced by their statutory powers (if any), the social and educational backgrounds and personalities of their members, their relationships with other opponents, their access to professional advice, their history, their performance in 1961-2 (if any), and their *raison d'être*.

Although the opponents did use traditional lobbying techniques in the second period, much of their time was devoted to technical tactics: developing alternatives to the Corporation's proposals, finding faults in Manchester's plans, and seeking safeguards to the schemes.

Many schemes were mooted in the press – Morecambe Bay and Solway Bay barrages, the Grand Contour Canal, desalination, cleaning Manchester's dirty rivers, and vague sites in the Pennines. But these were either impractical or were long-term projects. Manchester's organized opponents concentrated on two main sources as alternatives to the Corporation's Ullswater and Windermere proposals. These were the River Eamont and the River Leven at Greenodd. The River Eamont schemes were short lived, but there were numerous variations on the Greenodd schemes.

The Eamont scheme was abandoned for two main reasons: the Cumberland River Board opposed the scheme because it would have a greater effect than the Ullswater scheme on downstream flows and would complicate fisheries' schemes. In addition the consulting engineers for the opponents informed them that Manchester could expand the Eamont scheme as easily as the Ullswater scheme.

Two River Leven schemes were rejected by the opponents – one because it required a conspicuous reservoir; the second because the aqueduct in Longsleddale would have to be duplicated.

The Winster Valley Preservation Society favoured a scheme for abstraction from the River Leven at Greenodd and storage in Windermere. The Society believed that the scheme, by providing storage for all water in the Windermere catchment area, would make a Winster Valley reservoir unnecessary. It argued that the scheme would make the lake level more constant and hence would prevent flooding and improve recreation and fishing.

The fourth and most publicized Greenodd alternative was developed by Mr Jeffcoate, consulting engineer for the National Trust, the Council for the Preservation of Rural England, and other amenity organizations. Mr Jeffcoate proposed to abstract water from the Leven that would be treated and then passed into the Thirlmere and Haweswater aqueducts. Without storage the scheme would yield about 21 mgd. For 45 mgd, storage would be in a reservoir covering about 300 acres at Killington. This scheme could be expanded to provide storage capable of yielding 180 mgd.

The only alternative to the duplication of the Longsleddale aqueduct was suggested by a resident in Longsleddale (Mr T. D. Walshaw, Head of the Engineering Department at Harris College, Preston).

Many opponents saw their main hope of defeating Manchester in the discovery of a sound alternative scheme. It was believed that the Ministry would be favourably influenced if the opponents put forward alternatives at a Public Inquiry rather than merely protesting. Indeed the opponents were counselled by Mr W. Whitelaw, MP for Penrith and the Border, that the best way to defeat Manchester was at the Inquiry. It was his opinion that, if alternative schemes could be put forward that could produce the water that Manchester needed, then the Minister might be persuaded to turn down the draft Order.[17]

However, to suggest an alternative was not easy. As one opponent expressed it:

> It is simply no use putting forward alternatives which are going to be rejected as out of hand or alternatives which contain in themselves the same dangers as those we fear in the substantive case.

Certainly, opponents disliked each other's schemes. For example, the Windermere and Leven Valley Preservation Society and the National Trust believed that the Winster Society scheme for controlling Windermere would ruin the lake. On the other hand the Winster Society did not approve of the plan advocated by the National Trust and other organizations for Killington, as it just transferred Winster's problems to another valley.

In order to prevent a scheme being rejected without consideration it was thought necessary to have a professional water engineer defend the alternative. This produced many problems. Manchester had cornered almost half of the top water engineering firms in the country as its own consultants. Manchester's opponents had great difficulty in finding water engineering consultants who felt able to support any of their suggested schemes at the Public Inquiry.

The experts pointed out that it was difficult to make a case technically against abstraction from the lakes and in favour of abstraction from a river. Also they thought that the alternative schemes could not be completed in time. Only one water engineering consultant, Mr Jeffcoate, acting for the national amenity societies, was willing to appear at the Public Inquiry and argue for an alternative scheme on behalf of the opponents. Consultants and even planning officers were similarly reluctant to make technical criticisms of Manchester's plans. They believed that from an engineering point of view Manchester's proposals were the simplest, cheapest, and speediest method of meeting its short-term demand and that negligible damage would be caused to the lakes or rivers. One consultant pointed out that he was not an advocate and that it was not for the objectors to hand him a brief, expecting him to make out a case.

Although the engineering consultants were unable to fault Manchester's plans, they did urge their clients to seek safeguards. This generally entailed altering sections and details of Manchester's plans, ensuring that features important to the opponents were not omitted from the proposals, and making sure the provisions in the draft Order were interpreted correctly by both sides. Many authorities tried to negotiate safeguards before the Public Inquiry; others submitted requested safeguards along with their general objections to the draft Order.

In general, although many opponents concentrated on technical tactics, they had little success. It was difficult to find suitable alternatives and doubtful whether those alternatives that reached the Public Inquiry would be satisfactory to all bodies on either amenity or technical grounds. Few faults were found in Manchester's schemes, but a number of substantial safeguards were obtained.

Although much of the energy and hope of many of the opponents lay in technical opposition (alternative schemes, faults, and safeguards) some did pay attention to traditional pressure tactics – obtaining public support, lobbying the Government, and co-ordinating the opposition.

The opponents sought to gain public support by getting a good press, publishing newsletters, advertising, securing members and money, and holding public meetings.

As in 1961-2, the opponents were assisted by a sympathetic Lake District press. Not only did these papers give support in the second period, their staff also served as public relations officers to the Lake District preservation societies.

It is difficult to assess the success of the opponents' efforts to obtain public support in 1964-5. The local preservation societies were quite well supported but most of their members were the area's better educated, older, and wealthier inhabitants. The fact that the local preservation societies were led by rather aristocratic, wealthy individuals with a lot of spare time led some Lake District residents to state that they themselves were not 'preservationists'. It appears that there may have been a considerable body of opinion that was not in sympathy with what some residents call 'professional preservationists', 'out of touch millionaires' and 'Fiends of the Lake District'.

The statutory bodies and national amenity organizations did little political lobbying in 1964-5. In general, the local preservation societies confined their lobbying to joint efforts through the Lakes Defence Co-ordinating Committee and personal contacts of individual members.

The approach to lobbying was one area where the difference in views of the four local societies became marked. The Morecambe and Winster Societies were convinced that the way to win was to put forward a sound alternative to Manchester's proposals at the Public Inquiry. The Ullswater and Windermere Societies, on the other hand, placed more emphasis on political lobbying.

*Co-ordination of the opponents*

In theory, all opponents saw the desirability of co-operation and association. As Mr K. Dobell, Vice-Chairman of the Westmorland CC, expressed it: 'Everytime there appears to be a cleavage of opinion it gives tremendous encouragement to our opponents.'[18]

In practice, however, the opponents did not unite. As in 1961-2, bodies continued to be aware of the action of others mainly through overlapping membership, friendship networks, and shared staff. Geographical boundaries determined interests, and those in turn had some effect on how closely opponents thought, but had less effect on how closely the bodies associated. 'Association' in fact appeared to be decided on the basis of 'status'. In general, status is a function of legal authority, the size of the organization and its staff, and the age of the body.

Bodies on the same level associated freely even if they were not always in agreement. Those at lower levels attempted to work with those on a higher level but were, in general, rebuffed or patronized. As one preservation leader put it, the various types of opponents were of a 'different order'. Officials of County Councils made it clear that they could not commit themselves to working with voluntary associations.

Even though the national amenity organizations did present a joint case, little co-ordination was achieved between bodies of different status. In addition, the opponents had no uniting viewpoint. There was neither an effective co-ordinating principle nor a spokesman for all opponents in 1964-5.

Although groups of the same status associated with one another, the local preservation societies had difficulty working together. They did manage to combine in the Lakes Defence Co-ordinating Committee, for they realized the importance of unity. However, it is not much of an exaggeration to say that the differences between these groups were as great as the differences between them and Manchester. The societies disagreed over direct abstraction from the lakes, control of Windermere, and storage at Killington.

The four local societies were essentially parochial; the highest priority of each area was to preserve its own *status quo*. An unwillingness to compromise on this issue made it difficult to produce a unified policy. This, however, was complicated because each society had related principles: the lakes societies were against the development of any lake; the Winster Society decided that it could not transfer its problems to any other valley. Thus selfishness and altruism combined to eliminate most alternative schemes as any alternative would be unacceptable to at least one of the four societies. In deciding to condemn Manchester's plans to duplicate the Longsleddale aqueduct the societies removed about the only remaining area left for storage on which they could have agreed – Manchester's existing reservoirs.

Neither did the societies agree on the importance of presenting an alternative. To the Winster Society the only way to succeed was to suggest a practical alternative. The Society thought there was very little chance of preventing Manchester abstracting from Windermere and reasoned that it was necessary to present a scheme that would use all the potential of the Windermere catchment area to ensure that Manchester would not need the Winster Valley.

Mr Acland, North-West Area Agent of the National Trust, who held considerable sway over both the Windermere and Ullswater Societies, thought that the way to win was to try to keep Manchester out of the Lake District. Although the Ullswater and Windermere Societies did try to fight with alternatives, they were more willing to try other means as well. To the Winster Society, however, using other methods was a sign of defeat.

The skills and inclinations of the members of the four societies also differed. The members of the Winster and Morecambe Societies were practising and retired businessmen, a number of whom were engineers. The other societies largely consisted of landowning gentlemen with political connections. The Ullswater Society, and under the influence of joint members, the Windermere Society, were also conscious that they had won in 1961-2 by political opposition and some hoped that similar tactics might again succeed. The Winster Society however did not (as one of its members put it) 'suffer from the "hangover" of the successful opposition to Manchester's 1961 Ullswater proposals'. Its members also had a more favourable impression of water undertakings because of their business dealings. Not unnaturally all these differences were reflected in different policies.

All the societies were convinced that they had to be united to defeat Manchester and indeed some of the societies found limited co-operation easy. The lakes societies had similar interests, were equally threatened, and had common members. The Winster and Morecambe Societies also had similar interests (if either area were developed the other area would be affected), were both opposed to long-term plans, and had joint members. Beyond this unity, there was a general lack of sympathy with each other's cause. The Winster Society thought that at most those in the lakes societies would suffer a minor loss of amenity and deterioration of fishing rights, while Winster residents would lose their homes, livelihoods and land. In contrast, the lakes societies thought that they had immediate and real cause for concern and that there was no real threat to Winster; that the proposals for Winster were only put in to ensure concessions.

What is difficult to assess is whether the situation would have been worse had the Lakes Defence Co-ordinating Committee not existed. Probably much co-ordination would have taken place anyway through the overlapping membership on the societies. However, it is possible that, had these leaders not met every two weeks to talk over difficulties and to try to reach solutions, there would have been still more misunderstandings and even less co-operation. To the credit of the Co-ordinating Committee its many disputes remained internal and were not made public.

As the Public Inquiry approached, the opponents had had little success with their tactics to defeat Manchester. They had found alternatives but these were known to be costly, slow, and subject to many of the same criticisms as Manchester's plans. They had found few faults with Manchester's immediate proposals but had, by consultation, secured some safeguards. It is difficult to say whether their appeals to public opinion were successful. However, at one of its meetings the Lakes Defence Co-ordinating Committee was warned by its Public Relations Officer that there was no doubt the public generally was convinced that Manchester's proposals were benign, were acceptable to the majority and, therefore, the Committee could not expect the similar moral or pecuniary support given by the general public to the fight in 1962. Little lobbying was done. Although all the opponents realized the importance of unity, a unified front had not been possible. On this basis the opponents went to the Public Inquiry.

## The Public Inquiry

Great interest was aroused by the Public Inquiry into Manchester's draft Water Order. Of the seventy-five objections, forty-five were from 'public bodies, private companies, and individuals with a direct interest by virtue of landownership, riparian or other property interests, or authorities and others on whom notice has been served by Manchester';[19] the remaining thirty were non-statutory objectors, associations and individuals concerned primarily with the preservation of the National Parks, tourism, outdoor activities, and countryside beauty. Of the objectors, forty-six put their case at the Public Inquiry.

Manchester's claim that it needed more water urgently was generally conceded by the opponents.

The Corporation justified its proposals for Ullswater and Windermere by saying that they could be quickly implemented, would use very little agricultural land, utilized the surplus capacity in the aqueducts, and were approved by the Jellicoe Conference. They were also inexpensive and the country could not afford to spend more than necessary on the project, or to add to the cost of the goods which were produced using Manchester's water. The landscape would not suffer as great care would be taken with lake levels, most works would be underground, and the pumping stations would be noiseless. Access would still be available to the lakes. The duplicated aqueduct in Longsleddale was necessary in order to obtain the 25 mgd from Ullswater. Although some disturbance during construction could not be avoided, the whole project would be planned for minimal disturbance and the valley would be restored with meticulous care.

Opponents in general argued that such works should not be permitted in a National Park, and that Manchester's existing reservoirs had ruined what was previously unspoiled and beautiful. It was argued that, although buried, the pumping stations would be obvious by their access and the grass covering them might be a different colour from the surroundings. They doubted that no sound would emanate from the pumping station, and warned that it might be replaced with noisy pumps later. The lake level, the opponents anticipated, would be adversely affected by strong winds and possible drought Orders. It was feared that Manchester might later restrict access. Objections were also made to provisions for fisheries compensation.

Of the two lakes, most concern was expressed for Ullswater. However, the work which attracted most criticism was the Longsleddale aqueduct.

The opponents also expressed their fears for the future of the Winster Valley and especially of Ullswater, because of the surplus capacity in the proposed new Longsleddale aqueduct. Counsel for Manchester scoffed at this argument: 'That means that we are not to be allowed to do what is right today for fear of doing what is not right a second time.'[20] Manchester could do little to allay the opponents' fears for the future, but stated that the Council had not decided on future schemes.

The opponents argued that there were good and preferable alternatives to Ullswater and Windermere. A private objector suggested an engineering alternative to duplicating the Longsleddale aqueduct, but the emphasis was on Mr Jeffcoate's alternative (Greenodd-Killington) and to a lesser extent on the Winster Valley Preservation Society alternative (Greenodd-Windermere).

Those supporting Mr Jeffcoate's scheme maintained that the scheme was practical and that it had real advantages over Ullswater and Windermere. It used water from the River Leven at its mouth after all other users had benefited from the water; it was almost all outside the National Park; the only land to be flooded was agriculturally poor; and the scheme was capable of expansion.

The Winster Society scheme was said to be practical, to use water from the river mouth, and to be capable of expansion. It would flood no land, and by using all the water in the catchment area would eliminate the need for storage in the Winster Valley. Temporary supplies, it was suggested, could be borrowed from Liverpool. This would provide a valuable link and would allow more time for consideration of the alternatives.

In general, Manchester replied that all the alternatives needed much more careful consideration. A temporary supply from Liverpool would be unduly expensive; the water was of poorer quality; and the small surplus Liverpool now had would only last for a few years. It would also inhibit Liverpool from making much needed repairs.

Neither did the Corporation like the Leven-Killington scheme. It was said to be more costly on the grounds that the inferior water would require more treatment, the scheme was technically complex and the yield would fluctuate. Agricultural land would be used and as the aqueduct passed through the Winster Valley the valley would be threatened. Manchester presented letters from the Parish Councils in the Killington area to show that the scheme was generally disliked by the inhabitants.

The only real surprise or drama of the Public Inquiry came when the landscape consultant for the Corporation was cross-examined about his plans for Longsleddale. He agreed that if Manchester could obtain 20-25 mgd from other sources it would be a 'tragic folly' to interfere with Longsleddale, and that restoration of Longsleddale would be 'almost insuperable'. According to many this was the turning point of the Public Inquiry.

**The decision**

Following the Public Inquiry, many MPs repeatedly asked the Minister in the House when he would make his decision.[21]

The Inspector's Report was completed on 20 September 1965, and sent to Mr Richard Crossman, Minister of Housing and Local Government. The Report stated that Manchester needed more water and that the necessary water could be obtained from Ullswater and Windermere without damaging the amenities. However, the Longsleddale pipeline should not be allowed as it would cause 'permanent and irreparable damage'. Restrictions on abstraction from Ullswater, including a weir in the intake tunnel, were also suggested. The Inspector's view on alternatives was that they:[22]

> would need further study and detailed investigation to assess with certainty their feasibility. . . . They would be considerably more costly. . . . The works would be in beautiful country and the schemes would not be entirely free from objection.

He doubted if any alternatives could be completed before there was a grave risk of water shortage.[23]

The Minister's decision was announced on 6 May 1966. The decision-letter agreed with the Inspector's main recommendations. Manchester would be allowed to abstract water from Windermere and Ullswater, with restrictions on Ullswater abstraction, and would be forbidden to construct the Longsleddale pipeline. In addition to the Inspector's recommendations were the Minister's 'five additional concessions' to the amenity societies: that the works be subject to strict planning control, that Manchester submit loan applications to the Minister and that Manchester submit construction proposals to the Minister. The Minister stated that had it been legally feasible he would have stipulated in the Order that neither he nor his successors would allow Manchester to abstract more water from the lakes. He also insisted that Manchester prepare a programme for public access to Haweswater before he would make the Order.

It appears that the Minister was anxious to make additional concessions in order to ensure that the Order would not be opposed in Parliament. It was also necessary to placate opposition within the Cabinet from Messrs Marsh, Short, Willey and Peart.[24]

The Minister's belief that he had reached an admirable compromise was clear from his decision-letter. His modification of the Order, the omission of Longsleddale and the restrictions of abstraction from Ullswater:[25]

> would make the scheme less attractive from the water supply aspect in that the quantity of water produced might not be quite as much as Manchester were hoping to secure and the scheme would be less flexible for operational and maintenance purposes. But the proposals modified in these ways will, the Minister believes, do no harm to the Lake District provided sufficient care is taken over design, landscaping, etc.

The rest of the story is rather an anti-climax. Neither Manchester nor its opponents were delighted by the decision but they recognized it as a compromise. Manchester did not oppose because it wanted to start construction as soon as possible. Cumberland CC thought that, as the matter had been fully discussed at the Inquiry, and the decision was a compromise, it would be irresponsible for a public body to oppose. The Lake District Planning Board felt it might lose some of the Minister's safeguards by appealing. The national amenity organizations felt that an appeal would be expensive and unlikely to be successful. The Lakes Defence Co-ordinating Committee decided that its chairman, Lord Inglewood, should petition for a clause compelling Manchester to go back to Parliament in fifteen years for permission to continue abstraction from Ullswater. However, Lord Inglewood, and Mr Hall-Davis, MP, who supported this request, were both ruled not to have *locus standi* and the petitions were disallowed. No reasons were given. The only other organization to petition against the Order was the River Eden and District Fisheries Association.

Before this petition was heard Questions were asked in the House of Commons and there were debates in the House of Commons and in the House of Lords. However, there was no attempt to reverse the Order. The debates were general discussions which allowed the Lake District to repeat its fears for the future and which allowed Manchester to state its fear that owing to the Minister's safeguards it would soon be in urgent need of water once again.

The final hurdle for Manchester's Water Order was the petition of the River Eden and District Fisheries Association. The petitioners pleaded the fisheries case and requested safeguards. The Corporation strongly denied the need for such safeguards. Then, on the second day of the hearing, a compromise was arranged.

On 28 February 1967, the Government gave its final approval to the Water Order and announced that work could begin on 6 March 1967.

## Conclusions

In this dispute local preservation societies were formed to resist a threat and to preserve an area. However, although groups existed or were formed for Ullswater, Morecambe Bay and the Kent Estuary, the Winster Valley, and Windermere and the Leven, there were no preservation societies for Bannisdale, Longsleddale, or Killington. These three areas were also threatened and their inhabitants also opposed invasion. Why did they not form local preservation societies? None really considered the idea. There were few people in Longsleddale or Killington and fewer still in Bannisdale. Farmers in these areas were too busy looking after their land. In short, there were few people in these areas and those that were there were not what could be termed 'natural preservation society leaders' such as were to be found in the other threatened areas.

The leaders of the four local preservation societies in this controversy tended to be landowners and often leaders in the Country Landowners' Association, public school and /or university graduates, members of local authorities and /or pressure groups, self-employed or retired, financially well-off, middle-aged or older, and natives of the area. 'Leaders' here includes the official chairman and the very active significant members of the local preservation societies. Of these nine, six appear in Burke's *Peerage, Baronetage and Knightage*.

Persons with these characteristics, with one exception, did not live in Longsleddale, Killington or Bannisdale. Most of the residents of these areas were working farmers; many were small owner-occupiers.

Most theories of pressure group activity concentrate on the processes of pressure group activity – on the decision-making structure and on the characteristics of the pressure group. This dispute illustrates that both factors are significant. Certainly the opponents' membership in the House of Lords was decisive in the first period. Manchester's technical skills at the Public Inquiry were important in the second period. However, it is suggested that the real reason Manchester won was because of the substance of the dispute.

Most theorists place little emphasis on the subject matter of the dispute. Harry Eckstein in *Pressure Group Politics*[26] does say that the policy of the Government is one determinant of effectiveness. But in this dispute the government policy of ensuring an adequate water supply would appear to have been of overriding importance. It is suggested in this case-study that, because an abundant water supply is regarded as a basic necessity of life, a proposer of a new water scheme need only demonstrate that it is the only scheme that can be completed before demand exceeds supply in order to succeed. It has been demonstrated in this case-study that because of the nature of the subject matter of the dispute Manchester won and the opponents lost. Group characteristics and governmental structure can have some impact on a pressure group's success in securing safeguards but government policy and the nature of the dispute ensure that water undertakings win. Central to this theme is the idea that an adequate water supply must be provided. Also water engineering is sufficiently scientific to enable its practitioners to reach substantial agreement on supply, need, and the technical suitability of potential sources.

Manchester stated that it needed more water and produced schemes that could be completed before demand exceeded supply. Logically, the opponents must either show that Manchester does not need more water or find faults in Manchester's proposals and produce alternative schemes that could be completed before demand exceeded supply. The opponents tried to show that Manchester's need for water was exaggerated and failed; they tried to produce alternative schemes that could be completed in time and failed. The economic argument that money equals time and that alternative schemes could be completed in time by abandoning conventional engineering procedures and greatly increasing the men and money employed was never used. Opponents accepted their consulting engineers' estimates of time.

This idea and its consequences prompt important questions

relating to types of pressure and to alternative proposals. In this controversy it is obvious that, in the short-term, conventional pressure was used successfully by the opponents. Because the Lake District mobilized public support and lobbied peers, and because Manchester did little to enlist support from either the public or the Lords, the Lake District succeeded in throwing out the water clauses at the second reading of Manchester's Bill. Consequently, the Corporation was not given the opportunity of using technical arguments in a Select Committee in order to defend its proposals.

It was a commonly held view in the Lake District that while conventional techniques could affect a Bill, a more factual scientific approach was needed to influence the outcome of the draft Order. The general attitude was that the Inspector at the Public Inquiry would be an engineer who would reach his decision on the basis of technical evidence that was presented to him. This is to underestimate the role of the Inspector. He was at the Inquiry not solely as a judge with technical knowledge but as the Minister's representative. And it also ignores the fact that the final decision rests with the Minister, a politician as well as an administrator.

However, it is less easy to say what types of pressure are significant. In this case, Manchester won because of the superiority of its technical case. The opponents' case was based on both traditional and technical pressure. Accordingly, the opponents' success in obtaining concessions was a result of either traditional pressure, or attempts at technical pressure, or both. Although it is not possible to measure the contribution of these types of pressure, it appears that the Minister granted concessions because he feared that if the opponents opposed by traditional means in Parliament they would delay the implementation of the Order. Certainly the opponents' strong traditional pressure made the granting of political concessions more politically acceptable. From this study then, it appears that while successful technical pressure is of the greatest significance, traditional pressure may be effective in securing some concessions.

There is a general consensus that opponents should present alternative proposals. Certainly to suggest alternatives gives the opponents an aura of responsibility; they appear to be attempting to solve a problem in a constructive fashion rather than by merely negatively opposing. But are these alternatives very successful? In this dispute, although much work had been done on the alternative schemes, they were dismissed by the Inspector as being inferior on many counts to Manchester's proposals. More importantly they were rejected because they were incapable of supplying water in time.[27]

All alternative water-supply proposals suffer from the disadvantage that if the original proposals are defeated the alternative must then be formally promoted in a similar manner. In the case of a Bill, the promoters are required to petition for the changes and opponents are given the opportunity of objecting.[28] In the case of a draft Order, the applicant must publish notices of the change; if objections to the changes are not withdrawn another public inquiry must be held.[29] These procedures could occupy considerable time. As the time factor is usually significant, all alternatives are handicapped. Although it is believed by many (including, it is understood, Ministry officials) that an opponent in a technical case should suggest alternatives, the evidence of this study suggests that alternative schemes meet with little success. Alternatives may enhance an opponent's image but they are unlikely to defeat the original proposals. If this is generally true, it is of considerable significance where the provision of public utility and other essential services is contested on the grounds of the impairment of amenity.

**Editors' note**

The Ullswater-Windermere scheme is, at the time of writing, almost complete. Without an additional link between the Haweswater pumping station and the Watchgate treatment works, a serious situation could arise at times of peak demand. As a result, Manchester has submitted a draft Order for the construction of an underground Shap aqueduct. (For details see explanatory leaflet issued by Manchester Corporation, May 1972.)

## Notes

1 Much of the information for this case-study was obtained from interviews with participants and from their personal papers. I hope I have not offended anyone by minimizing their efforts in this controversy, in my attempt to be brief and diplomatic.
2 Ministry of Health, *The Cheshire and Lancashire Water Survey* 1949.
3 Press release, 1 September 1961.
4 *Guardian*, 19 December 1962.
5 Lord Birkett was a famous barrister, former judge at Nuremberg, a native of the Lake District and a former President of the Friends of the Lake District.
6 Lord Lonsdale owned considerable land near Ullswater and was a member of the Ullswater Preservation Society, the National Trust and the Country Landowners' Association.
7 *Parliamentary Debates* (Lords), 8 February 1962, col. 233.
8 Ibid.

9 Ibid., col. 237.
10 Ibid., col. 345.
11 Ibid., col. 344.
12 Ibid., cols 351-2.
13 *Conference on Water Resources in the North-West*, HMSO, 1963, para. 30.
14 *Report on Future Water Resources*, Manchester Corporation Waterworks, January 1964.
15 Letter from Mr P. J. Liddell to *The Times*, 30 March 1964.
16 Letter from Chairman of the Morecambe Bay and Kent Estuary Preservation Society to the *Westmorland Gazette*, 21 February 1964.
17 *Cumberland and Westmorland Herald*, 8 May 1965.
18 *Penrith Observer*, 4 May 1965.
19 Ministry of Housing and Local Government, *Abstraction of Water from Ullswater and Windermere* (the Inspector's Report), 20 September 1965.
20 Recording of the Public Inquiry.
21 *Parliamentary Debates* (Commons), 29 October 1965, 2 November 1965, 3 November 1965, 13 December 1965.
22 Inspector's Report.
23 Ibid.
24 The *Sunday Times*, 13 March 1966, and the *Manchester Evening News*, 24 April 1966.
25 The Minister's decision-letter, 5 May 1966, para. 25.
26 Harry Eckstein, *Pressure Group Politics*, Allen & Unwin, 1960.
27 This conclusion differs from that reached by Roy Gregory in *The Price of Amenity*. Gregory has stated 'typically, the judgement becomes a matter of deciding whether an intangible amenity, with no determinable money value, is worth a specific price. That price is the difference between the cost of the project under application and that of the possibility generally regarded as the most likely alternative' (p. 33). In this dispute, although this type of argument was made, the significant factor was not money but time. The financial cost of the alternative schemes was not the significant factor; the significant factor was that an adequate water supply was regarded as essential and the alternatives were incapable of producing water before demand exceeded supply.
28 Barnett Cocks (ed.), *Sir Thomas Erskine May's Treatise on the Law, Privileges, Proceedings, and Usage of Parliament*, Butterworth, 1964, 17th ed.
29 Water Act 1945, sched. I, part I, paras 6 and 7.

# Chapter 6
# The Minister's line: or, the M4 comes to Berkshire

**Roy Gregory**

The music of Sibelius is appreciated in Britain because it is like the British Government: reticent and slow of delivery.

Sir Thomas Beecham

**Drawing the line**

Maidenhead in Berkshire and Tormarton in Gloucestershire are nearly seventy miles apart as the crow flies, and between them lies some of the most attractive countryside in the south of England. Finding the most suitable and acceptable route for the London-South Wales motorway between these two points has proved to be a difficult and delicate exercise, and one which admirably illustrates the type of problem that arises when administrators are obliged to take into account not only complicated technical and financial factors but also intangible and unquantifiable aesthetic considerations.

The number of feasible lines for a motorway between any two points may be large, but it is not infinite, and sooner or later a choice has to be made among the two or three possibilities that are obviously more promising than the rest. With aesthetic controversies superimposed upon the usual conflict of material interests, and with the Ministry of Transport committed to a policy of securing the widest measure of acceptance for its proposals, disputes of this kind are likely to be both heated and protracted. When the motorway programme was set in motion in the 1950s the target was 1,000 miles of new road in use by the early 1970s. By the end of 1965, 375 miles had been completed, 141 miles were under construction, and another 536 miles were at various stages in the planning pipeline. Progress has not been faster for a variety of reasons, among them lack of money, and a statutory procedure of formidable length and intricacy. Problems like that of finding the most economical and acceptable line for the Maidenhead-Tormarton section of the M4 have made their own special contribution.

**The procedural obstacle course**

Any student of British government who claims to detect a disquieting trend towards administrative ruthlessness and arrogance in the interests of speed and efficiency and at the expense of the rights and property of the individual should draw comfort from an examination of the procedure through which the Minister of Transport goes when planning the route of a motorway and acquiring the land necessary for its construction. The Minister's powers and duties are laid down in the Highways Act of 1959, an enormously long statute which consolidates a great deal of earlier legislation, including the Special Roads Act of 1949. In the earlier stages of the planning process the main objective is for the Minister to make a scheme, under Section 11 of the Highways Act, fixing the line of the motorway. Once a scheme has been made the land required can be protected against development for any other purpose. On occasions, a suitable line will already exist on a county council's approved development plan, and the Minister may then simply follow this route. Where there is no suitable line, and the procedure has to begin from scratch, making a scheme may be a much more protracted affair.

The responsibility for motorway projects is sometimes entrusted by the Minister to county councils (more particularly the larger ones) on an agency basis; on other occasions, where it seems to the Minister that a stretch of motorway would be beyond the scope of a county's staff, he appoints consulting engineers to carry out surveys and recommend the most promising and feasible lines from which a final choice can be made. By law, the Minister is bound to 'give due consideration to the requirements of local and national planning, including the requirements of agriculture', and so it is customary, at this stage, to consult informally all interested local and national bodies in order to try to secure their agreement to the line preferred by the Ministry. This may involve lengthy discussions with, for example, the Ministry of Agriculture, Fisheries and Food, the Ministry of Housing and Local Government, the Ministry of Power, nationalized industries, local authorities likely to be affected, the National Farmers' Union, the County Landowners' Association, and national and local amenity societies. Individual property owners are also given every opportunity of stating their views to the Ministry at this stage. When the Minister is finally satisfied that he has the best line he can get, this is published as a draft scheme. It is advertised in the press, and the government departments, local authorities and public utilities affected are notified. Publication of the draft scheme is required even when the line has been previously included in a local authority's development plan. There follows a period of three months during which objections may be lodged by individuals or public authorities. If there are objections, the Ministry will then enter into discussions with the objectors, and will try to meet their wishes so far as possible in the hope of persuading them to withdraw their objections. If this fails it may be necessary to hold a local public inquiry. Under the terms of the 1959 Act, the Minister must hold an inquiry if there is an unresolved objection from a county borough, a county council, or county district through which any part of the motorway runs, or from a navigation authority where a bridge is involved; where the objection comes from any other quarter the Minister may hold an inquiry at his discretion. If there is an inquiry an independent inspector is appointed, and a period of at least three months is required for arrangements to be made and for the interested parties to prepare their cases. After considering the inspector's report the Minister may decide to go ahead and make the scheme unmodified, or make it with modifications, or not make it at all. He need not accept the inspector's recommendations, but if he does not he must state the reasons for his decision. If he decides to make substantial changes or abandon the original draft scheme, the whole procedure for settling the line must start again at the beginning.

A new motorway will almost certainly involve alterations to side roads. They may have to be fed into the motorway, taken over or under it, or stopped up altogether, and to make alterations of this kind the Minister requires statutory authority. Under Section 13 of the Highways Act draft orders have to be made for these changes, there is a three-month period for objections from local authorities, other public bodies and individuals, and under certain circumstances there may have to be a public inquiry.

After the scheme settling the line of the motorway and the orders affecting the side roads have been made, there follows a period of six weeks during which anyone still aggrieved by the Minister's proposals may apply to the High Court and challenge the validity of the scheme or the orders. It is then open to the Court to make an interim order suspending operations until the case has been decided. And if, on hearing the case, the Court rules that the scheme or orders were not made within the powers given by the Act, or that the applicant's interests have been substantially prejudiced by the Minister's failure to comply with the provisions of the Act, it may quash the scheme or orders *in toto* or in so far as they affect the property of the applicant. The mere possibility that a scheme may be challenged on legal grounds, thereby holding up progress for a year or more, whatever the outcome of the case, naturally leads the Ministry to deal circumspectly with objectors, so as to leave no opening for the plea that the Minister has not given careful consideration to their views.

In the second phase of the planning process the Ministry has to acquire the land necessary for the new road. When the statutory position is assured and the design of the motorway and side roads is sufficiently advanced to show the limits of the land that will have to be acquired, negotiations can begin with the owners or occupiers of the plots that are needed. If a particular plot cannot be acquired by agreement, the Minister resorts to the compulsory purchase procedure. In practice it is necessary to seek compulsory powers of acquisition where major schemes are concerned to avoid complications when the contractors need to move on to the land required. A draft compulsory purchase order is served on the owner or occupier of the plot needed, there is a three-week period for objections, and if there are objections and they cannot be resolved, an inquiry must be held. In this event an inspector is appointed, and the Minister must consider his report before deciding whether to make the order with or without modification, or whether to make it at all. It may well take the Ministry a year or more to acquire all the land required for a major scheme, and even longer if difficult problems arise in connexion with rehousing residents, or moving business or industrial undertakings on the route. When right of entry on to the land has been secured the Minister can invite tenders from selected contractors. The choice of tenders is made, the contracts are let, and at last building the motorway can begin. It comes as no surprise to learn that whilst the actual construction of the M1 from London to Rugby took twenty months, the preparatory stages occupied eight years. The signs are that this ratio will be beaten by the Berkshire section of the M4.[1]

### The problem emerges

It was in 1938, as part of a national road system proposed by the County Surveyors' Society, that a new motorway between London and the south-west first came under serious consideration. At that time, and for many years afterwards, the line envisaged for the motorway followed a route rather similar to that of the A4, the existing trunk road to Bath, and really amounted to a series of linked by-passes, aligned south of Slough, Maidenhead, Reading, Newbury and Marlborough. In May 1946 the Government outlined a ten-year programme for roads, and included among the tentative plans for 800 miles of motorway was the proposal for a new road from London to South Wales. Soon afterwards, the Ministry of Transport asked the Berkshire County Surveyor to investigate a possible line from Maidenhead to pass south of Reading and Newbury. In 1949 came the Special Roads Act, empowering the Minister of Transport to make schemes for the provision of motorways, and in 1956 it was announced that five major projects were to be given priority. These were: improving the Great North Road, a motorway from the Midlands to South Wales, and motorways from London to the north-west (via Birmingham), to the south-east ports, and to South Wales. Whether or not a motorway from London to South Wales really rated such a high priority must remain a moot point; at all events, the decision had now been taken, and planning a route for the 133 miles of the M4 could begin in earnest.

The practice followed by the Ministry of Transport when fixing the line of a motorway is to divide it into a number of sections, making a separate scheme for each section. Two sections of the M4, the Maidenhead and Slough by-passes, presented no difficulty for here the Minister felt able to adopt lines that had been settled before the war, and which had subsequently appeared on the Buckinghamshire and Berkshire county development plans. A scheme for these two stretches, covering between them eleven miles of the route, was made as early as February 1957.[2] The twelve-mile section between Langley, at the eastern end of the Slough by-pass, and Chiswick in west London, did prove more of a problem, because the Ministry of Transport decided to take the motorway out of London on an elevated viaduct over two miles in length, the line of which departed from the ground level route laid down in the Middlesex development plan. The Middlesex County Council needed a good deal of convincing of the wisdom of this change, and there were also strong objections from individuals and firms affected by the viaduct. In October 1960, however, after a public inquiry held in the summer, the Minister made a scheme for this section, and now the line was fixed for the entire twenty-three miles from London as far west as Maidenhead. At the other end of the route, too, reasonable progress was being made, and in September 1961 the Minister published a draft scheme for an eleven-mile section of the M4 in Gloucestershire between Tormarton and Almondsbury.[3] Thus there remained a central section, from seventy to eighty miles long, between Maidenhead and Tormarton for which the line still had to be settled.

One possible route, of course, already existed. The line proposed in 1938, and generally referred to as the Bath road route, left the Maidenhead by-pass just east of Maidenhead Thicket, swung south of Reading, continued south of Newbury, through Savernake Forest and south of Marlborough. In fact, the first thirteen miles of this route, from Maidenhead Thicket to a point due south of Reading at Three Mile Cross, appeared in the 1958 review of the Berkshire county development plan. Though the development plan even now has not yet been finally approved by the Ministry of Housing and Local Government, it was being generally assumed in Berkshire and Reading at this time that eventually the extension of the M4 would follow this line. But when the Ministry of Transport began to turn its attention to the Maidenhead-Tormarton section in the late 1950s it soon became apparent that there would be a number of serious drawbacks to the Bath road route.

**Towards a decision**

The Ministry's view is that the primary purpose of a motorway is to provide a fast, direct and safe through route for long distance traffic between the major centres of population and industry. Other things being equal, the most direct line is always to be preferred because the shorter the route the lower the continuing costs to users, the lower the initial cost of construction (at this time three-lane dual carriageway motorways through rural areas were costing on average between £600,000 and £700,000 a mile to build), and the lower the annual bill for maintenance and repairs. The first objection to the Bath road route was that it was certainly not the shortest that could be devised. The second purpose of a motorway is to serve the cities and more important towns en route by taking the line, where possible, to a distance of between two and five miles from these places. The obvious objection to the Bath road line on this score was that, whilst it might serve Reading, it ran much too far south to be of any use to Swindon. The Ministry had other subsidiary reasons, too, for disliking the old Bath road line: there would be serious engineering difficulties in trying to build a motorway in the low lying and badly drained area to the south-west of Reading; the road would have to cut through Savernake Forest; it would attract local traffic that ought to stay on the A4; and it would have to climb to a considerable height in order to cross the Marlborough Downs. Much of this was plain enough even before November 1960 when Mr Marples announced that he would soon be appointing consulting engineers to carry out surveys.

The consulting engineers began work in the spring of 1961, and by the end of the year they had produced a preliminary report which tended to confirm the Ministry's dislike of the Bath road route. The consulting engineers had also been instructed to look at alternatives, and during the winter of 1961-2 their views on other possible lines were examined in the Ministry. In April 1962 Mr Marples announced that he would soon be consulting organizations representing agriculture and the local authorities concerned. Three lines now seemed to be worth considering. There was the original Bath road route, which could not yet be entirely dismissed; there was a much more northerly route through the Vale of White Horse and north of Swindon; and there was what subsequently came to be called the Direct route which ran north of Reading and then virtually straight over the Berkshire Downs. The Bath road route was 79 miles, and the Vale of White Horse route 77 miles long. The Bath road route served Reading and not Swindon, and the Vale route served Swindon and not Reading. The Direct route was only 72 miles long, or about 10 per cent shorter than either of the other two, and into the bargain it served both Reading and Swindon. In the Ministry's view, therefore, the Direct route, or something like it, was clearly preferable to either of the alternatives. But even in these early days, while opinion in the Ministry was hardening in favour of the Direct line, a certain amount of opposition to this route was beginning to develop.

In 1956 the Ministry of Transport had set up an Advisory Committee on the Landscape Treatment of Trunk Roads. It arrived upon the scene too late to be consulted about the lines of the earlier motorways, and at first had to confine itself to making suggestions for the cosmetic treatment of roads, the routes for which had already been determined on engineering grounds. By 1959, however, it was inspecting proposed lines and giving its views on the effect that any particular route would have on the local landscape and amenities. Though the advice given to the Minister by the Committee is, of course, confidential, there is reason to think that the Committee was not unanimous in its recommendation and that there was some opposition to the Direct route. Among the members of the Advisory Committee was the chairman of the Council for the Preservation of Rural England, Mr (now Sir) George Langley-Taylor and he certainly preferred the Bath road line.

Also represented on the Committee was the Ministry of Housing and Local Government, and this Department had views of its own about the most suitable route for the M4. The Ministry of Housing and Local Government, with its general responsibilities for land-use and the development of new towns, naturally approached the problem of selecting a motorway line from a standpoint rather different from that of the Ministry of Transport. To its way of thinking, the line should have been chosen with a view to serving not only existing industrial centres, but also potential growth points in between. It agreed that the line would have to run much closer to Swindon than the old Bath road route; but it also argued that the route ought to take account of potential growth in the Newbury and Hungerford areas. The South East Study, proposing a new city based upon Newbury, was not to be published until February 1964, but evidently the possibilities of this area had already attracted its attention. There was the further question of land-use within the area: if one substantial part of the region was to be scheduled for industrial and residential development, it was essential that the unspoiled stretches of countryside that remained should be preserved for recreational purposes.

The Ministry of Housing and Local Government therefore proposed a route which passed south of Reading and Newbury and then turned north-west along the line of the A419 passing west of Aldbourne to join the Ministry of Transport's Direct route near Liddington. This line, they claimed, would serve Reading, Newbury and Swindon, and at the same time would avoid the Berkshire Downs. After a good deal of argument it was rejected by the Ministry of Transport partly because of engineering difficulties in the vicinity of Aldbourne, but mainly because of its excessive length. In any case, the Ministry of Transport tends to believe that town-planners consistently over-estimate the volume of external traffic generated by new towns, and for this reason it was not much swayed by the argument that a major deviation from the most direct line ought to be made in order to serve a possible new town to which, as yet, there was no definite commitment from the Government.

## Mr Marples shows his hand

Nothing of this, of course, was as yet public knowledge. But surveyors at work cannot be concealed, and their arrival on the Downs in the autumn of 1961 naturally aroused the suspicion and alarm of the local landowners. They in turn alerted others who were prepared to fight to preserve this stretch of countryside, and in April 1962 a group of such people, judging that the Minister would soon show his hand, fired the first public shot in a controversy that was still flickering nearly five years later. On 27 April 1962, above the names of an impressive list of artists, architects, writers, MPs, and other public figures, which include John Betjeman, Ruth Dalton, Lord Esher, Allen Lane, Frederick Etchells, Rosamund Lehmann, Lawrence Gowing, Lord Moyne, Lord Norris, Barnet Stross and Woodrow Wyatt, there appeared in *The Times* a letter deploring the mere idea of a motorway over the Berkshire Downs. They dwelt at length on the beauties of the higher contours of the Downs and the horrors of a motorway. Setting the pattern for almost everyone else who was to object to one or other of the many proposed routes, they went on to give the Minister the benefit of their advice upon which other parts of the countryside the new road should damage, and which other people it should disturb. Their view was that the M4, like the line of the GWR more than a hundred years earlier, ought to follow the suitably level Vale of White Horse. They ended with the first of the many ringing appeals that were to be made to the Minister's finer feelings in the course of the next few years: 'We feel, in short, that the engineers in the Ministry of Transport should not be allowed to choose a route which, while it may offer advantages of convenience and low compensation, will annihilate for ever one of the fairest parts of the nation's heritage of beauty.'

Soon afterwards, in May 1962, the Minister opened informal consultations with the major local authorities affected, namely Berkshire, Gloucestershire, Hampshire, Oxfordshire and Wiltshire, the County Borough of Reading and the Borough of Swindon. They were asked for their reactions to each of the three routes and were left in no doubt that the Minister himself had a 'broad preference' for the shortest and most direct of them. The Ministry's policy is to go to very great lengths to secure the widest measure of agreement from local authorities which will be affected by the line of a trunk road, and this is especially true of those local authorities which can oblige the Minister to hold a public inquiry if they refuse to withdraw their objections. But even the best prepared schemes can run into unexpected hazards, and at this point it is worth making a short digression to describe a particularly unhappy experience connected with another section of the M4, which the Ministry was reflecting upon at just this time in the spring of 1962.

## A cautionary tale

The proposed route for the Tormarton-Almondsbury section of the London-South Wales motorway had first been made public in the Gloucestershire county development plan in 1951, and subsequently the line itself had been protected against development. But in 1953 the County Council allowed houses to be built quite close to the line at Downend along a 2½-mile stretch, part of which fell within the Warmley Rural District. In 1951 Warmley had approved the plan in principle, and as late as 1960 it offered no objection to the quinquennial review of the plan, upon which the motorway line still appeared. In September 1961 the Minister published a draft scheme for this section, and on 31 October a letter was received from the Warmley Rural District Council saying that it had no objection. But on 30 November, one month later, and ten days before the objection period expired, the Ministry received another letter: now the Warmley Rural District Council did have an objection. Some of the residents of Downend had evidently discovered that the noise and disturbance caused by a motorway extended farther than they had realized, and had persuaded the Warmley Rural District Council to object on the grounds that the new road would spoil their amenities. This story came to light in an adjournment debate in the House of Commons in March 1962 when the Parliamentary Secretary to the Ministry of Transport, Mr John Hay, dealt with the charge made by the MP for South Gloucestershire, Mr F. V. Corfield, that these residents had been harshly treated. Judging by the tone of Mr Hay's speech the Ministry had been astonished and infuriated by this turn of events. Warmley refused to withdraw its objection, and so there had to be a public inquiry, which was eventually held in August 1962. 'This case bids fair', said Mr Hay, 'to become a classic in the annals in the Ministry of Transport. If ever it were necessary to show the sort of frustrations and difficulties that we are up against in trying to get the motorway network built this is that case.'

What particularly disturbed the Ministry was that this stretch of motorway connected with the Severn bridge section, and the bridge, at a cost of £16 million, was due to be completed in 1966. If the Tormarton-Almondsbury section were to be held up for any length of time, the chances were that the bridge would be finished while a large part of the road leading to it would be still under construction. 'I must warn my Hon. Friend (Corfield)', said the exasperated Mr Hay, 'that I would not be surprised if many motorists, the RAC and the AA and many others in the Bristol area had something to say to the Warmley Rural District Council for slowing up this matter. . . . We shall give the Warmley council and the residents of Warmley every opportunity to put their case, but it is our clear intention that we want to get the motorway built. . . . I must warn him that the risk is that we will now be seriously delayed in carrying through this important and expensive scheme because one rural district council, covering a couple of miles, feels that, having agreed all the way through since 1951, it must now change its tune and have a public inquiry.'[4] These were strong words, prompted perhaps by the certain knowledge that if the bridge was finished before the approaches to it were ready, it would be the Ministry of Transport and not the Warmley Rural District which would take the blame.

## A county divided against itself

With this episode very fresh in mind, and indeed with the public inquiry still pending, the Minister awaited the views of the local authorities affected by the Maidenhead-Tormarton section. It was their reaction to the Direct route that chiefly interested the Ministry, and it was no doubt pleased to learn that Swindon, Wiltshire, Gloucestershire and Hampshire had no objection to it. The Berkshire County Council had no objection either, which on the face of things was rather more surprising, in view of the passionate opposition to the Direct route among many individuals in the county. The responsibility for collecting the views of the district councils, boroughs, and other interested bodies within Berkshire had been handed over to a joint committee made up of the Planning and Highways and Bridges committees. According to the local press, opinion among the local authorities in Berkshire was fairly evenly divided: one group, composed chiefly of those authorities which would not be affected by it, favoured the Minister's Direct route, and the other group either opposed this line or expressed a preference for the old Bath road route. The joint committee had also taken note of the views of a number of other organizations: the Ramblers' Association opposed the Direct route, the Faringdon branch of the National Farmers' Union opposed the Vale route, and the Berkshire branch of the Council for the Preservation of Rural England favoured the Bath road line.

The first of the many local amenity societies that were to be in action over the next few years had also made their views known. There was a Vale of White Horse Preservation Society, opposing the Vale route on the grounds that it would pass through heavily populated countryside, thereby causing considerable disturbance, that there would be exceptionally difficult drainage problems in the Vale, and that fog was prevalent. Furthermore, despite the prevalence of fog, there was a magnificent view from Uffington Hill, which would be spoiled by a motorway. There was a Downs Preservation Committee, opposing the Direct route, on the grounds that it would spoil a beautiful stretch of countryside, that it would interfere with the racehorse gallops in the vicinity of Lambourn, and that in winter ice and high cross-winds would make the road dangerous, as well as entailing expensive maintenance costs. There was a Kennet Valley Preservation Society, opposing the Bath road route on the grounds that it would divide and damage a great many smallholdings in the valley. Faced with these conflicting views, the joint committee recommended that the Council should offer no objection to the Direct route, and this recommendation was accepted by the full Council when it met in July 1962. All those who objected very strongly to the Direct route naturally resented what they called the 'spineless' attitude of the Berkshire County Council. But the truth of the matter was that the County Council was in a very difficult position. As a major local authority, it felt that it would not be acting in a responsible way if it simply opposed the Ministry's proposal without offering some positive alternative which the county would be prepared to support. Since opinion within the county was so divided, there would have been just as much opposition to any other route as to the Direct route, and consequently there was no chance of the County Council uniting behind a firm counter-proposal. In the circumstances there seemed to be no way out but to say nothing and let the Ministry of Transport take the responsibility for a decision that was bound to arouse a great deal of ill feeling.

## 'And he smelleth the battle afar off'

So far, the Minister's Direct route had enjoyed a deceptively comfortable passage. However, the two remaining local authorities concerned, Reading and Oxfordshire, were both vehemently opposed to it. Reading was greatly alarmed by the prospect of a motorway running north of the town, for it felt that most of the traffic generated in the growing industrial area in the southern part of the borough, quite apart from general traffic in the centre of the town, would try to join the new road by travelling northwards through Reading itself. The two bridges under the east-west railway line, and the two bridges over the Thames, already constituted serious bottle-necks for north-south traffic, and if even more traffic started to flow in this direction, the Borough Council feared that the congestion would become intolerable. Sooner or later there would have to be an extensive programme of road improvements to cope with this problem, and this would not only cause a great deal of unnecessary disturbance, but would also involve the Borough in very considerable expense.[5] The Oxfordshire County Council was also against the Direct route, and could see no reason why the original Bath road line should be abandoned; the Chairman of the council, Lord Macclesfield, remarked that it looked very much as though the Minister had simply told the consulting engineers to take the shortest line from A to B and 'see what happens'. It was a shrewd enough observation, for the Ministry of Transport never really expected that the Direct route as proposed in May 1962 would prove acceptable in its entirety and without modification in detail. There might be little chance of the Direct line emerging totally unscathed; but in the Ministry's view, if public attention could be focused on this new line, and diverted from the unsuitable and excessively long southern route, that in itself would be a considerable gain.

From this point the Direct route started to run into strong opposition from other quarters, too. The Minister was now to discover that the Berkshire Downs have many powerful defenders, and at this stage they began to make their voices heard. In March 1962 there had been some correspondence between the Ministry and the National Parks Commission[6] about the possible effect of the M4 on stretches of countryside which were likely to be designated areas of 'outstanding natural beauty'. The Commission, under the chairmanship of Lord Strang, a former Permanent Secretary at the Foreign Office, is a corporate body with the statutory duty of making representations to Ministers on any proposed development likely to be prejudicial to the countryside in England and Wales. After the Ministry had put forward its three routes in May they were inspected by the Commission's Field Adviser and some of its members, and on the basis of their reports the Commission informed the Ministry of Transport that the Direct route would be most undesirable from a landscape point of view, and that the Bath road route was much to be preferred.

The National Parks Commissioners were not the only people taking an interest in the Downs in the summer of 1962. The Royal Fine Art Commission, a prestigious body under the chairmanship of Lord Bridges, a former Permanent Secretary at the Treasury, had also been officially informed of the Minister's proposal for a Direct route over the Downs. The Commission is empowered by royal warrant to inquire into and call attention to any development which in its opinion might affect amenities of a national or public character. Some of its members, like Lord Bridges, John Betjeman, Mr G. A. Jellicoe and John Piper were already familiar with the Downs and had personal connexions with this part of the country. When they looked at the possible routes in August, they too informed the Minister that they objected to the Direct route and that the Bath road line would be preferable. Another inspection of the Direct route had been carried out by Mr George Langley-Taylor, the Chairman of the Council for the Preservation of Rural England, on behalf of the Executive of that body. Although the Council is a voluntary and non-official organization, its views are thought to have considerable influence in matters affecting the countryside. In June 1962 its Executive accepted Mr Langley-Taylor's recommendation that they should oppose both the Vale and Direct routes, but subject to detailed survey they should raise no objection to the Bath road route.

By this time there was also another local amenity society in the field. Soon after the Minister's views had been put to the local authorities affected in May 1962, a Chilterns and South Oxfordshire Preservation Society had been formed for the purpose of defending the area through which the eastern section of the Direct line would pass. This was to prove a particularly active and articulate society, and it went to a great deal of trouble to build up a case against the Direct route. By July 1962 it had set out a number of arguments against the Direct route in a memorandum submitted to Mr John Hay in his capacity as the local MP for Henley. It claimed that the gradients on the Direct route were such that the travelling time for loaded lorries was longer, and therefore haulage costs would be higher than on the admittedly longer Bath road route;[7] that the motorway would meet adverse weather conditions in the Oxfordshire Chilterns; that the cost of two additional bridges over the Thames, plus a feeder road to south Reading, would outweigh any savings in length of road; that the fencing necessary to keep deer off the road in the Chilterns would add considerably to the cost; and that nothing could justify ruining areas of outstanding natural beauty at Shiplake, in the Goring Gap and in the Chiltern beechwoods when alternative routes were available.

There can be little doubt that by this time the Ministry of Transport had no intention of adopting the Bath road or the Vale of White Horse routes, however much their merits might be urged by local authorities and amenity bodies. In principle, the Direct route still had everything to recommend it. Yet there was no denying that a case had been made against the line as proposed in May 1962. As it then stood, the Direct route would undoubtedly have damaged the Oxfordshire beechwoods, it would have impaired the facilities for racehorse training, and it would certainly have entailed what the Parliamentary Secretary was later to describe as a 'pretty big and difficult structure' in the Goring Gap. There was no question of abandoning the idea of a short route to the north of Reading and over the Downs. But there was no reason why the original Direct route should not be revised in an effort to meet objections. The consulting engineers were therefore asked to look again, and at the end of November 1962, in answer to a Question in the House about the progress of the Minister's deliberations, Mr Hay could only reply that a careful study of all the views expressed was still going on. 'Nothing gets people in this country more excited', he said, 'than the possibility of natural countryside being despoiled. That is what has caused delay.'

## The Minister tries again

On 23 January 1963 the Minister made public a fresh proposal. This new line, which was put forward as a 'basis for consideration', and which was to be known as the Revised Direct route, ran to the south of the earlier Direct line crossing the Thames just south of Shiplake and just north of Tilehurst. From the Ministry of Transport's point of view the Revised Direct route had a great deal to recommend it. It was fractionally shorter than the Direct route, and in comparison with the original Bath road line the Ministry estimated that it would save the traffic using it £1½ million a year in running costs, besides reducing the annual maintenance bill by about £17,500. Furthermore, the Ministry seems to have thought that this new line had been devised in a way that was sufficiently ingenious to meet most of the objections that had been raised against the Direct route over the Downs. Among the advantages claimed for the Revised Direct route were that it was closer to Reading, that apart from one short stretch it avoided the high downland, that the western crossing over the Thames could be made inconspicuous because the area was heavily wooded, that it avoided the Oxfordshire beechwoods, and that it avoided most of the racehorse gallops on the Downs. When the new proposal was published it was accompanied by a brisk announcement that detailed surveys would now begin and that the Minister hoped to be able to publish a draft scheme by the end of 1963.[8] The signs are that the Ministry was trying to convey the impression that this was the best line so far produced and that it was not likely to be shifted from it. In February, in a speech in his constituency, Mr Hay reaffirmed that the draft scheme was expected to be ready by the end of the year, though, as he pointed out, the attitude of the local authorities affected would govern the speed with which it could be made. 'If nobody objects', he said, 'it could be quick.' More realistically, he added: 'We are not very optimistic.'

In fact, as the Ministry was soon to discover, the Revised Direct route had even more enemies than the Direct route which it replaced. None of those who had opposed the Direct route accepted the revised version; some of them, indeed, liked it even less. And at the same time, this new proposal stirred up fresh opponents where there had been none before. From now on all the objectors were to fight harder still, some because their sense of danger was heightened by the suspicion that the Minister was more firmly committed to this line than to any earlier proposal, and some because the knowledge that he had shifted his ground once under pressure encouraged them in the belief that if they protested loudly enough he might give way again.

Having published the new proposal, the Ministry's next step was to repeat the earlier procedure and consult the local authorities concerned. Once again, and for reasons already described, the Berkshire County Council proved to be agreeably compliant, and on the recommendation of the joint Planning and Highways and Bridges Committee, which as before had been given the job of canvassing opinion in the county, it agreed to tell the Minister that it had no objection in principle to the Revised Direct route provided that special attention was paid to the Thames crossings and the line across the Lambourn valley. This is not the place to explore the internal politics of the Berkshire County Council, but its attitude first to the Direct route and now to the Revised Direct route was provoking a great deal of discontent. Many of the council's members were readily prepared to fall in with the Minister's proposals simply in order to bring to an end, as quickly as possible, the delay that was causing so much uncertainty in the county. At the same time others, who for one reason or another bitterly opposed the two routes so far suggested by the Minister, were pointing out with some heat that the Minister, with the endorsement of the council, was rejecting a line south of Reading which the council itself had planned. There was also a suggestion that the southern part of the county had more, and perhaps more influential representation on the council, and that these representatives were not going to go out of their way to persuade the Minister to bring the motorway back on to their own doorsteps when he seemed so determined to take it elsewhere.

The Berkshire County Council, however, was almost alone in seeing no objection to the Revised Direct route. From Reading's point of view the Minister's second thoughts were if anything even more objectionable than his first. The new line still ran north of the town, and would therefore bring just as many traffic problems as the Direct route. Furthermore, since it ran closer to the northern boundary of Reading – this was an advantage in the eyes of the Ministry – it might seriously hamper the borough's future expansion northwards. If the motorway had to go north of the town it was in Reading's interest that it should go as far north as possible.

Neither were the national bodies concerned with the preservation of amenity impressed with the Revised Direct route. At the end of November 1962 the Minister had informed the National Parks Commission and the Royal Fine Art Commission of the proposals he intended to make for a modified version of the Direct route. The National Parks Commission informed the Minister that his new favourite was not much better than the old, and the Royal Fine Art Commission condemned it out of hand. The Council for the Preservation of Rural England shared their objections to the Revised Direct route and was also thinking about putting forward a positive alternative of its own. As a member of the Landscape Advisory Committee Mr Langley-Taylor was familiar with the amended version of the Bath road route which had been urged upon the Ministry of Transport by the Ministry of Housing and Local Government. Believing that if this were revived in a slightly different form, it might provide a feasible alternative to the Revised Direct route, he devised a new line, similar to that suggested by the Ministry of Housing and Local Government, except that it ran to the east and not the west of Aldbourne. This proposal, which was to be known as the Revised Bath road route, was to be put into circulation in May 1963.

In the meantime, it was becoming clear that none of the Minister's former critics in Berkshire and Oxfordshire had been appeased by the Revised Direct route. The Downs Preservation Committee conceded that the new line was slightly better than the old, but they still wanted a route south of Reading. The Chilterns and South Oxfordshire Preservation Society also rejected the Revised Direct route, and by March 1963 it had assembled enough new ammunition for another memorandum to be submitted to Mr Hay. And now that a new line was under consideration, two entirely fresh sets of residents were alarmed, and two more preservation societies sprang into existence. In moving the line from the top of the Downs the Minister had brought it much closer to a string of villages running along the southern slopes of the hills, and in February 1963, predictably enough, a Berkshire Downs Villages Association was formed to fight the Revised Direct route. By July another group of property owners, between Reading and Maidenhead, had seen the red light and set themselves up as the Mid-Thames Valley Association.

From all of these quarters the Ministry was now to be bombarded with a formidable battery of arguments, in the shape of deputations, letters, memoranda, and specially commissioned reports. The main arguments advanced against the Revised Direct route were that the cost of constructing two bridges over the Thames and a southern by-pass round Reading would outweigh any savings that might result from taking a shorter route; that the undulating nature of the Revised Direct route would increase the travelling time of loaded lorries to such an extent that hauliers would incur greater costs than on the flatter Bath road route; that where the road crossed the Thames flood-plains at Shiplake and Mapledurham costly drainage works would have to be undertaken; and that in order to take the motorway from the lower northern bank of the river in Oxfordshire to the higher southern bank in Berkshire, and at the same time clear an existing road and railway, an elaborate viaduct would be needed at the Tilehurst crossing. And finally, of course, there was Reading's argument that any route north of the town would bring insoluble traffic problems to the borough.

### . . . And sticks to his guns

But for the time being, at least, there was no sign that these arguments were making any impression upon the Ministry. In April

1963 Mr Hay was interviewed at his home in Silchester and took the opportunity of dealing with some of the criticisms that had been levelled at the Revised Direct route. The Ministry had allowed for the cost of two bridges over the Thames, he said, and their route would still be cheaper. The Bath road route south of Reading would have to cross numerous streams and brooks, the Kennet and Avon Canal, and a large number of minor roads, and this would entail a great many small bridges, the total cost of which would exceed that of two bridges over the Thames. Because the land south of Reading was low lying and subject to flooding, construction work would take longer and be more costly. All in all, the average cost per mile on the Revised Direct route would be £700,000 whilst on the Bath road route it would be £1 million per mile.

Turning to Reading's case, Mr Hay now proceeded to stand the borough's argument on its head. The Ministry had been advised, he said, that no less than 32 per cent of Reading's industry was north of the railway line. If the route ran south of Reading, traffic from the industrial premises in the north of the town would have to pass under one of the two railway bridges, both of which had low headroom, and neither of which could be easily improved. Furthermore, the roads running southwards out of Reading were substandard and were not amenable to improvement and widening without the substantial demolition of property. If the motorway passed to the north of Reading, however, traffic from north of the railway line would not have to pass through the centre of the town to reach it. What traffic there was from south of the town would be able to join the motorway at interchanges planned to the west and east of Reading at Twyford and Purley. Even if Mr Hay's figures had been correct, which they most certainly were not, this was a curious argument, for it implied that to plan in such a way as to prevent one-third of Reading's traffic from travelling through the town centre from north to south was more important than avoiding the costly arrangements that would be necessary to prevent two-thirds from moving along the same roads, in the opposite direction. Soon after this Mr Hay departed from the Ministry of Transport to become Civil Lord of the Admiralty, and was replaced by Mr T. G. D. Galbraith.

## The guardians of amenity in arms

The Ministry of Transport's obvious predilection for a route north of Reading had no doubt come as a considerable relief to those property owners and residents in the Kennet Valley who would have been disturbed if the original Bath road line had remained in favour. Whilst there was nothing as yet to suggest that the Minister was likely to be deflected from his Revised Direct route, the force of the campaign against this latest proposal was beginning to cause a certain amount of uneasiness in the Kennet Valley, and in June 1963 the Kennet Valley Preservation Society set up an Action committee to defend the valley and to remind the Minister of the excellent reasons why he should not consider any route south of Reading. The Kennet Valley residents soon had more cause for alarm. On 30 July the Council for the Preservation of Rural England organized a conference at the Royal Society of Arts building in London. All the many bodies, local and national, that were interested in this section of the M4 were invited to send representatives and it seems to have been a lively meeting. The chairman of the CPRE, Mr Langley-Taylor, publicly revealed for the first time the Revised Bath road route which his Council had decided to promote. Naturally, all those who objected to the Revised Direct route were well pleased with this new line. But one branch of the CPRE, the Wiltshire branch, strongly opposed their Council's proposals, and the two representatives from the Kennet Valley were very disturbed indeed by what they saw on the CPRE's wall map. When they returned to Berkshire with their report immediate steps were taken to turn the Kennet Valley Preservation Society, hitherto a rather shadowy body, into a much more effective organization, which by January 1964 could claim to have over five hundred members and a large fighting fund.[9] On the advice of the Civic Trust, it commissioned two independent experts, Mr R. W. Rose, the County Surveyor and Planning Officer for the Isle of Wight and a qualified civil engineer, landscape architect and town planner, and Mr D. Rigby Childs, a planning consultant, to produce a report on the merits of the routes so far proposed. As the summer wore on there was more evidence that the defenders of the Kennet Valley had cause for anxiety, for in August 1963 there were reports that the surveyors had been seen at work on a line that passed south of Reading and Newbury, crossed the Kennet east of Kintbury, and then turned north-west to pass east of Aldbourne. This looked suspiciously similar to the CPRE'S Revised Bath road route.

At the same time, the opponents of the Revised Direct route were also growing apprehensive. On 16 August it was reported in the *Daily Telegraph* that Mr Marples had told Sir Anthony Hurd that he personally still favoured the amended Direct route across the Downs. And even if the consulting engineers were looking at the Revised Bath road route, there was reason to believe that the only line being surveyed in full detail was the Revised Direct route. Some of those who objected to this line were beginning to reason that once the Minister had committed himself to the extent of publishing a draft scheme the chances of a major alteration would be very slim. Rightly or wrongly, it was suspected that the Minister might now be quite close to publishing a draft scheme. The time had come, therefore, to bring the whole issue to the notice of a wider public, and for two weeks, between 10 and 26 September the scene of the controversy shifted to the correspondence columns of *The Times*, whose readership was treated to the spectacle of two former Permanent Secretaries, both chairmen of quasi-governmental commissions, roundly condemning a Minister's proposals. Lord Bridges for the Royal Fine Art Commission, Lord Strang for the National Parks Commission and Mr Langley-Taylor for the CPRE all wrote attacking the Revised Direct route and urging the case for the Bath road line.

The Ministry of Transport was reluctant to admit that it was wavering. In October 1963, however, Mr Galbraith conceded that because of the highly controversial nature of the route the consulting engineers had been asked to acquire information about possible alternatives before making their final report, although, as he put it, 'the Minister's decision to select the Revised Direct route for a detailed survey remains completely unchanged'.

It will be remembered that in January 1963 the Ministry had hoped to publish a draft scheme before the end of the year: in a written answer to a Parliamentary Question the Minister was obliged to say on 11 December 1963 that he did not expect to receive the consulting engineers' final report before the spring of 1964 at the earliest.

The new year saw the controversy still in full swing. In January the Kennet Valley Preservation Society published the survey undertaken for it by Messrs Rose and Rigby Childs. Of all the many memoranda and reports that were produced this was by far the most costly and elaborate, running to eighty-two pages, and lavishly illustrated with maps, diagrams and photographs. According to Mr Rose, if certain amendments were made to the Revised Direct route it would have a less adverse effect on the countryside than the Revised Bath road route. He claimed that the large-scale countryside through which the western section of the Revised Direct route would run could easily contain a motorway, whilst the smaller-scale countryside of the Kennet Valley could not: that there were greater opportunities for the Revised Direct route to follow the natural ground levels: that the Revised Bath road route would have to cut into a series of hills between Chilton Foliat and Aldbourne in a way that would disrupt the character of that area; and that the Revised Direct route would affect a smaller area of great landscape value. Up to this point those who opposed a route south of Reading and through the Kennet Valley had always had to base their arguments upon considerations of cost, length, construction difficulties and disturbance. Now they had authoritative support for their views on amenity grounds, too. To his main report, Mr Rose added an interesting coda which was to turn out to be of great importance. If the Minister were finally to decide that the motorway must go south of Reading it could still avoid the Kennet Valley by means of what Mr Rose called a 'link road'. This would swing north-west from a point south of Reading at Three Mile Cross to join the line already proposed for the Revised Direct route, thus avoiding both the western section of the Bath road route, and the admittedly difficult eastern section of the Revised Direct route.

The publication of this report produced an immediate and predictable response in Berkshire. The Berkshire Downs Villages Association, the Downs Preservation Committee, the Mid-Thames Valley Association and the Chilterns and South Oxfordshire Preservation Society all rapidly expressed their total disagreement with its findings. The Downs Villages Association announced that it would soon be presenting the report of its own independent expert, with the implication that it would certainly reach a very different conclusion.

## A pause for reflection

It may well have been that at this point the issue was very much in the balance, with the odds still slightly in favour of the Revised Direct route north of Reading. But in February 1964 there were fresh developments which obliged the Ministry of Transport to think again, and which probably started to turn the tide against the Revised Direct line. First, there was the publication of the results of the Reading traffic survey. This survey was begun early in 1962 when the Ministries of Transport and Housing and Local Government had required the borough council to support its proposals for the redevelopment of the town centre and for an inner ring road by evidence from a traffic survey. Reading decided to carry out a comprehensive land-use/transportation survey, designed in such a way that the information derived from it could be used for a number of purposes, one of which was to predict the effect on the town of the various routes proposed for the M4. At first, the Ministry of Transport showed little interest in this aspect of the survey; but in April 1963 there was a change of heart, and the Ministry asked if Reading would supply the results of the survey for processing along with other data being collected. Reading agreed to this request, but because by this time the borough's limited staff was fully occupied in analysing the results in support of the town centre proposals, it was not until November 1963 that it was able to pass certain basic information, without written comment or explanation, to the consulting engineers. In February 1964 when Reading made public the results of the survey, they told heavily against the Revised Direct route. The figures showed that nearly 90 per cent of Reading's traffic came from or went to an area south of the east-west railway line. Armed with these figures, which, of course, completely demolished those put forward earlier by Mr Hay, the Reading Council could now argue that if the motorway was built to the north of the town, the volume of traffic trying to pass through in a north-south direction would certainly entail so much alteration to the centre of Reading that it would more than cancel out any gain resulting from the construction of a shorter motorway. The borough council was now prepared to support the link road idea as proposed by Mr Rose in his report to the Kennet Valley Preservation Society, for this route, so it was felt, would attract a good deal of the Wokingham to Wallingford traffic, thus significantly reducing the weight of traffic passing through Reading itself. (It was exactly local traffic of this kind that the Ministry of Transport was anxious to keep off the motorway.)

Second, also in February 1964, there was published the *South East Study* 1961-1981, envisaging three new cities in the south-east. One of them was to be based on Newbury, the population of which was to rise from 20,000 in 1961 to 95,000 in 1981, eventually reaching more than 150,000 or even a quarter of a million. If a development of this magnitude really was to take place in the Newbury area, there was a case for at least looking again at the line of the motorway that was to serve it. And third, in February 1964, Mr Marples announced that the British and French Governments had decided in principle to go ahead with the Channel Tunnel project. The volume of new traffic likely to be generated by the tunnel is not large; nevertheless this was another new factor that had to be taken into the reckoning. In any case, the final report from the consulting engineers was taking longer than expected, and in answer to a Question in the House in February the Minister admitted that it now looked like being the autumn of 1964 before he could reach a decision.

When Mr Peter Emery, the MP for Reading, raised the question of the route of the M4 on the adjournment on 25 March, the debate came at a particularly awkward moment for the Ministry of Transport, for it was still weighing up the implications of the developments that had occurred in the previous month. The Ministry must have been aware for some years past that there was a possibility of a new town in the vicinity of Newbury; and there had been serious talk of a Channel Tunnel from at least as far back as 1957. But talk by civil servants is one thing and a public commitment on the part of the Government is another. Consequently, extra weight now had to be given to these two considerations. The results of the Reading traffic survey had been received at the Ministry only on 4 March, and so when the debate took place there had been very little time in which to digest them.

For these reasons, when the Parliamentary Secretary, Mr T. G. D. Galbraith, put the Ministry's view he had no alternative but to go on repeating all the familiar arguments in favour of the Revised Direct route as though they were as conclusive as ever. Mr Emery, who had naturally been carefully briefed by the officials of the Reading Corporation, pointed out that for some years planning and development in the town had taken place on the assumption that the motorway would follow the line laid down on the Berkshire development plan; earlier proposals for a loop road within Reading, linking the Oxford road with the London and Wokingham roads had therefore been abandoned. That route had been subsequently built over, and now it would be very difficult to provide any kind of loop road. If the Minister insisted on fixing the line of the M4 north of Reading then he would have to ask him to guarantee a road running from Calcot on the west of Reading, round the south of the town, to the Sonning and Twyford area on the east: this, he estimated, would be about ten miles long and would cost £425,000 a mile.

In reply, Mr Galbraith could not add much that was new, but he did indicate that the Ministry's initial reaction to the newly proposed link road was not particularly favourable. Construction costs on both the link road route and the Revised Direct route were broadly similar, despite the need for two bridges over the Thames on the latter. Since the link route was nearly three miles longer such advantages as there were on the score of cost lay with the Revised Direct route. It seemed to the Ministry, he said, that Reading in its anxiety to use the motorway as part of a relief ring road to keep north-south traffic out of the town, was taking a view of the functions of a motorway that was diametrically opposed to that of the Minister. And whilst it was understandable that Reading should want to use the motorway to help solve its own traffic problems, he was not sure if this was a legitimate use of a main national route to the west.[10] If there were a great many short-distance drivers entering and leaving the motorway, this would obviously cause unnecessary hazards to the long-distance traffic for which it had been designed. The Ministry, he said, had always recognized Reading's difficulties, and its preliminary view was that traffic from the industrial areas to the south of the town would use the proposed junctions to the east and west of Reading. Even if Mr Emery's figures for the cost of a relief road were correct, the financial advantages of the Revised

Direct route would still not disappear because of the savings on mileage that would accrue to through traffic. This was clearly a holding operation on the part of the Parliamentary Secretary, for he concluded his speech with the remark that nothing he had said should be interpreted outside the House to mean that the Minister had made up his mind.

### 'Fight the good fight . . .'

In Berkshire there was no let up in the battle of words between the preservation societies. The Berkshire Downs Villages Association was now ready with its impartial report. The expert commissioned was the landscape architect, Mr G. A. Jellicoe, a member of the Royal Fine Art Commission, which had already condemned the Revised Direct route. It was not surprising that Mr Jellicoe too should turn out to be vehemently opposed to this line. In his view it would do 'monstrous damage' at the Goring Gap; it would break unnecessarily into new territory, and even if it could not be seen, it would still destroy the sense of repose; it would cut across ground-modelling of the most sensitive kind; it would be like 'a circular saw cutting across the fingers of the hand that is controlling it'. He dwelt at length upon the horrors of a motorway: 'The proposed motorway would not only be contrary to the natural process but would infest the area with all the urban diseases of which motorways are both the heirs and the carriers.' It has already been pointed out that those who opposed one route usually felt it incumbent upon themselves to urge upon the Minister the merits of an alternative. To describe the motorway in Mr Jellicoe's terms in one breath, and in the next to insist, as he did, that it ought to follow a line south of Reading, was not likely to endear him to the defenders of the Kennet Valley. Neither was the next step in his argument calculated to please them, though the point was valid enough: since a large part of Berkshire was bound to become urban in character, and this was particularly true of the Kennet Valley, they must ensure that an outlet was provided for the inhabitants of the new towns to enjoy the real countryside. They must therefore preserve that valuable stretch of countryside where the Revised Direct route ran. In other words, since the Kennet Valley was doomed anyway, it might as well have the monstrous motorway, too. When it is a question of preserving the countryside passions run high, and it seems that there is no barrel so deep that it cannot be scraped: Mr Jellicoe's last point was that on the steeper gradients of the Revised Direct route diesel powered vehicles would emit more toxic fumes which would reduce visibility and make overtaking more dangerous.

**Where we came in . . . ?**

The summer of 1964 wore on, there was still no word from the Ministry of Transport, and in Berkshire and Oxfordshire suspense and anxiety were giving way to sheer exasperation at the continuing uncertainty.[11] As it happened, the Ministry had also been thrown into a position of some uncertainty at this point. In May the consulting engineers' final report had arrived, and there were some disquieting deductions to be drawn from it. So unexpectedly rapid had been the natural growth in traffic between 1960 and 1964 that it now seemed certain that when the Tormarton-Maidenhead section of the M4 was completed the extra traffic consequently generated would very soon have the effect of overloading the existing Slough and Maidenhead by-passes, which, of course, had not been designed to modern motorway standards and have only two-lane carriageways. This was a new and alarming element in the situation, and clearly at some none too distant date in the future a second motorway would be required between London and the Reading area to relieve the pressure on the two by-passes. The Reading traffic survey had already weakened the case for a line north of Reading. There was now an even more powerful argument in favour of a route south of the borough, for if it was going to be necessary to think in terms of another motorway running on a line to the south of the Slough and Maidenhead by-passes, it would obviously make more sense to bring the M4 from Tormarton to a point south of Reading from which the relief motorway could be extended to connect with new roads leading out of south London. Furthermore, if in the very long run there were to be any substantial flow of traffic from South Wales in the direction of the Channel Tunnel, a motorway terminating at Chiswick in west London would not be of much use to it.

And to cap all, the Ministry could now see a way of taking the M4 south of Reading without involving itself in all the difficulties associated with the earlier proposals for a southern route. The clue had been provided by the Kennet Valley Preservation Society report: the line could pass south of Reading, then swing north-west and at a point to the west of the town continue along a path similar to that of the Revised Direct route, which all along had had a great deal to recommend it in the eyes of the Ministry. Naturally, a reappraisal of this kind did not take place in five minutes, and in July in answer to Questions in the House, the Parliamentary Secretary let it be known that the consulting engineers' report was still being studied and that the autumn was the earliest likely date for the publication of the draft scheme.[12]

The autumn of 1964 brought a General Election, a new Government, and a new Minister of Transport in the person of Mr Tom Fraser.[13] It did not bring a draft scheme for the Maidenhead-Tormarton section of the M4. In November, when Mr Fraser replied to Questions about the delay he disclosed that the Ministry had found it necessary to refer back for further consideration a number of points in the consulting engineers' final report. By January 1965 there were rumours of rethinking in the Ministry of Transport, and in February for those who could read between the lines, there were further hints that something new was in the offing. In answer to the now familiar Questions about delays and dates the Minister announced that he hoped to be able to publish a draft scheme for the thirty-mile Tormarton-Liddington section of the M4 in the following month: this was something, although there had never been much controversy about this stretch. The Maidenhead-Liddington section, he said, presented more difficulty because since the consulting engineers had carried out their original survey there had been certain developments of wide significance: there had been the South East Study, there had been the information from the Reading traffic survey, and there was now the prospect of a Channel Tunnel. In the circumstances he had asked the consulting engineers to re-examine certain sections of their report, and had also appointed another firm of consulting engineers to carry out a study of the approaches to the M4 from the south of London and from the extreme south-east of England.

At long last, in April 1965 the Ministry was able to place before the Reading and Berkshire Councils a highly confidential document giving details of a new proposal and asking for their reactions to it by the end of May. It seems to have been almost impossible to keep these confidential communications from the Ministry completely secret, and it was not long before fairly detailed descriptions of its contents were in circulation in Berkshire. Soon the news was out that the latest line was to go south of Reading. The wheel had come full circle.

For two months the Ministry digested local reaction. Then, early in August 1965 details were announced of the route upon which the Minister had finally settled, and which he intended to publish in a draft scheme. This final line, forty-seven miles in length (i.e. about five miles longer than the Direct or Revised Direct routes between Maidenhead and Liddington), was to leave the Maidenhead by-pass near Holyport, swing south of Reading, and then head north-west between Reading and Theale. From a point north of Theale it was to run west on a line rather to the south of the earlier Revised Direct route, passing south of Lambourn and along the Roman Ermine Street to Liddington in Wiltshire. The draft scheme itself was not expected for another six months, and then it would be only for the nineteen-mile stretch between Maidenhead and Theale. The draft scheme for the remainder between Theale and Liddington, could not be expected until the middle of 1966.[14]

During the period when the Ministry was in retreat from the Revised Direct route a good deal had been heard about the Reading traffic survey, the South East Study and the Channel Tunnel. In the statement accompanying the details of the new line, however, the Ministry chose to stress the attention that had been paid to amenity considerations. And now, of course, the Minister was at pains to draw attention to the advantages of a route south of Reading: it would avoid the necessity for two expensive crossings over the Thames, with their associated approaches and viaducts; it would be able to take advantage, for the first thirteen miles from Maidenhead, of a line that had been protected for many years in the Berkshire development plan; the route between Shinfield and Theale crossed an area that was being extensively worked for gravel, and therefore the effect upon amenities would be small; by passing to the south of Reading the motorway would serve the industrial areas that were mainly on that side of the town, and at the same time would draw off a good deal of the traffic that passed through the town on the A4.

In August 1965 the Minister had arrived at almost precisely the same point in the procedural obstacle course as he had reached in January 1963; he had made his informal proposals and stated his intention of publishing a draft scheme, based upon these proposals, at some later date. But in 1965 all the signs were that the Minister had finally succeeded in picking a way through the Berkshire and Oxfordshire minefield. With the exception of Oxfordshire, all the major local authorities, Reading, Berkshire and Wiltshire, were satisfied and, apart from the Berkshire Downs Villages Association, all the local amenity societies which had opposed a route north of Reading were pleased; the main opponent of a route to the south of Reading, the Kennet Valley Preservation Society, was prepared to describe the line as the 'best of a bad job', for, though it did run south of Reading, it turned north-west before reaching the main part of the Society's bailiwick. The National Parks Commission, the Royal Fine Art Commission and the Council for the Preservation of Rural England were none of them entirely happy with the line to the west of Reading, but were no longer determined to oppose the latest route in principle. To this extent the story has a happy ending. Yet, whilst it is obviously desirable and prudent for a Minister to try to avoid giving offence to those most immediately affected by his policies and decisions, his first duty is to provide the whole community with value for money. There should certainly be no suspicion, however unjustified, that the construction of an important new road has been unnecessarily delayed by a prolonged search for the line of least resistance or by excessively zealous efforts to conciliate sectional and local interests.

**Right answer, wrong working?**

No one outside the Ministry of Transport possesses enough information to reach an independent verdict on the merits of the final choice as compared with the claims of the several other routes that were considered. Comment upon the way in which a decision was reached is another matter, and here the Ministry of Transport is clearly open to a number of criticisms.

In the first place, the Ministry took a very long time to make up its mind. In answer to a Question in the House of Lords in May 1965, Lord Lindgren, the Joint Parliamentary Secretary, pointed out that discussions about the route of the M4 had been in progress since 1950 and remarked, apparently in all seriousness, that no one could complain of undue haste in taking a decision. This is certainly the last accusation which anyone would wish to level at the Ministry. Even if it is accepted that a serious effort to find the most suitable line for the Maidenhead-Tormarton section did not begin until 1961 when the consulting engineers were appointed, it still took five years to publish a draft scheme for one stretch of this section, and publishing a draft scheme, it will be recalled, is a fairly early step in the administrative procedure. Impressive statistics have been produced to demonstrate the economic benefits of motorways; since none of the benefits come into play until the motorway is in use, substantial costs must be imputed to any unwarranted delay that occurs during the planning stages. As it has taken so long to establish the line of the Maidenhead-Tormarton section, it is possible that at some future date there will be completed stretches of motorway at either end of the route, connected by the existing all-purpose trunk roads. Motorways both attract and generate traffic, and if experience elsewhere is any guide, these existing trunk roads are likely to become heavily overloaded, even to the extent of causing delays and congestion on the motorway sections themselves.[15]

To arguments of this kind the Ministry might reply that the right decision is of more importance than a quick answer. Once built, a motorway will be there for many years, and time spent on finding the most satisfactory line is time well spent when measured against the total life-span of a motorway. And at first sight, in the case of the Maidenhead-Tormarton section of the M4, this might seem to be a particularly appropriate reply, for if the Ministry believes, as presumably it does, that the route proposed in August 1965 is the most suitable, then an earlier decision in favour of any other line would necessarily have produced what would now be considered a less satisfactory route.

This answer, however, would merely leave the way open for a second and more important criticism; and this is that with more foresight and imagination, and with more accurate forecasting, the Ministry would never have wasted its time on the earlier proposals, and need never have been in danger of settling for a route that would have proved to be unsatisfactory had it been adopted. In other words, the Ministry should never have given serious consideration to the Direct route and Revised Direct route, and should have seen the necessity for something akin to the eventual solution at a much earlier stage. There would certainly be some substance in these criticisms if, but only if, there were reason to believe that in 1961 or 1962 the Ministry ought to have possessed the information, or been able to make the predictions, which finally persuaded it to accept a route south of Reading, and ought to have been able to gauge more accurately the weight of opposition to its various proposals for routes north of Reading.

In one context or another the Ministry ascribed its change of heart between January 1963 and August 1965 to four factors. These were: (i) the Government's acceptance of the South East Study proposals; (ii) the Government's commitment to a Channel Tunnel; (iii) new information from the Reading traffic survey; and (iv) objections on amenity grounds to the Direct route and Revised Direct route. The first two considerations, the alleged implications of the South East Study and the Channel Tunnel, may be dismissed at once, for they can have had relatively little effect on the final decision; in fact, it is difficult to see why they were given such prominence, unless it was to draw attention away from other and more important grounds for the reappraisal.

Should the Ministry have been better informed at an earlier date of the effect upon Reading of a motorway north of the town? It was, of course, only by chance that the facts of the situation came to light when they did, or at all, for Reading had undertaken its traffic survey primarily for reasons that had no connexion with the M4, and there are no grounds for thinking that such a survey would have been carried out had the borough not been seeking approval for its redevelopment scheme. It would be easy to condemn the Ministry on the score of inadequate preparatory work, for it could be argued that a central department ought to look into the local side-effects that might be set up by its own decisions, particularly when the consequences of those decisions might be far-reaching enough to raise questions about the use of national resources. On the other hand, it could equally well be maintained – and this is indeed the view of the Ministry of Transport – that the responsibility for assessing and demonstrating the nature and extent of any local side-effects rests primarily with those local authorities who claim that they will be adversely affected. True, there will be times when the local consequences of a department's decisions turn out to be important enough to persuade the department to think again. But any consequences that are economically significant in terms of national resources will, *a fortiori*, be of even greater significance, proportionately, in terms of local resources. And given that there are local authorities charged with the duty of protecting local interests, it is at least arguable that it is they who should investigate and assess the consequences for themselves of departmental policies, so as to improve their chances of influencing the Minister in the course of the consultations that he is statutorily bound to hold with them.

Neither can it be said with any confidence that the Ministry ought to have foreseen the outcry that its Direct and Revised Direct routes would arouse on amenity grounds. There were competent judges, including some on the Landscape Advisory Committee, who genuinely believed that rolling and open countryside like the Berkshire Downs could easily absorb a motorway. And, in any case, since there is no objective way of quantifying amenity benefits, it may be that their value can be assessed only by flying trial balloons and observing public reaction. This, of course, is exactly what the Ministry did, though at the beginning of the episode it probably did not altogether appreciate what a hornet's nest the M4 was plunging into beyond Maidenhead. Furthermore, there are many practical advantages in giving a full and fair hearing to local authorities, amenity societies, and individual property owners who may be affected, even if this does spin out the consultative process to great lengths. The more information that the Ministry has, the better the chances of reaching the most appropriate decision. The variety and complexity of the considerations that may be relevant to fixing a motorway route need no emphasizing, and with the best will in the world, administrators and engineers may easily overlook or underestimate difficulties and objections that are much better appreciated by people on the spot, whether they are local councillors, farmers or preservationists.[16] The simple exercise of looking again, and meeting objections with reasoned arguments obliges the Ministry to check its facts and figures and reconsider the logic of its own case. Naturally many of those who ply the Minister with objections based ostensibly on amenity grounds are motivated primarily by a quite understandable desire to protect their own self-interest. But the rules of the game are well understood: the Minister will listen only to evidence that is couched in terms of the public interest, and the knowledge that this is so obliges any objector who wishes to be taken seriously to restrict himself to arguments and data that may be genuinely useful in reaching a decision. At the end of the day, the Ministry settled for a route very similar indeed to a line suggested by an expert commissioned by one of the local amenity societies.

One last reason for the Ministry's change of heart remains to be considered. The balance was finally tipped in favour of a route south of Reading when the Ministry realized that the Maidenhead and Slough by-passes were in danger of becoming overloaded, and that it ought to be planning for a relief motorway running in the direction of south London. Had the rate of growth in the volume of traffic in the early 1960s been more accurately predicted at an earlier date, this factor might have been taken into account much sooner. Yet it is by no means certain that on its own the eventual need for a second motorway would have been a decisive argument, for a spur from the Revised Direct route passing south of Reading and on to London would have been perfectly feasible. It was only *in combination with* the other arguments, the force of which did not become apparent until a later stage, that the necessity for a new road in the future proved conclusive.

It is true that in a sense it was the element of chance that saved the Ministry of Transport. Had the nature of the countryside between Maidenhead and Liddington been different, and had the districts through which the various lines passed contained fewer persistent, determined and articulate individuals, an earlier – and therefore less satisfactory – planning decision might well have been reached. Yet when every stage in the story is examined in detail it is hard to see what the Ministry could have done that it did not do, and even harder to argue that it did anything that it ought not to have done. And when the formidable complexities of the problem are unravelled, the Ministry can scarcely be grudged its share of what luck was going. With the retrospective wisdom that comes readily to academic snipers it would be easy enough to dismiss the Ministry's performance as short-sighted and unimaginative. The real lesson of this story may be that busy administrators are of necessity so preoccupied with the pressing day-to-day problems on their desks that there is not enough time for long looks and wide views.

## Notes

1 It was in fact opened in December 1971.

2 The Maidenhead by-pass had been first proposed in the early 1920s; the line was fixed in 1936 and the land acquired in 1938. The surface soil for the road had actually been excavated by the outbreak of the war. After

the war the Ministry of Transport decided to stick to the original design which meant that the by-passes had two rather than three lanes in each direction. This turned out to be a decision of some importance.

3 As it happened, a last minute objection from a local authority through which the route ran obliged the Minister to hold a public inquiry, and so delayed part of the scheme for a considerable time.

4 Eventually the Ministry of Transport split the Almondsbury-Tormarton section of the M4 into two parts. A scheme for one stretch, between Almondsbury and Hambrook, which was not affected by Warmley's objection, was made in April 1963. After the public inquiry on the other stretch, from Hambrook to Tormarton, in August 1962, a modified line was published in a draft scheme in May 1963 and a scheme made in September 1963. This section of eight miles was opened in December 1966.

5 *Reading Standard*, 6 July 1962. It is true that Reading would have received a grant towards the cost of improving its two main north-south roads, one of which was class I (subject to 75 per cent grant) and the other class II (subject to 60 per cent grant). Even so, the borough would still have been required to meet about a third of the total bill.

6 Now the Countryside Commission.

7 Among the members of this Society was a prominent local road haulier; the Society was therefore able to conduct practical experiments to back its claims about the effects of gradients on the length of journeys.

8 *Reading Standard*, 25 January 1963. The Minister seems to have been quite sensitive to the possible charge that he had been unduly influenced by the 'racehorse fraternity', and in letters to Peter Emery (MP for Reading) and Sir Anthony Hurd (MP for Newbury) went out of his way to emphasize that their views had not played any major part in his decision. *Reading Mercury*, 23 June 1963.

9 All the routes considered seriously by the Ministry of Transport would have passed close to or through the property of prominent public figures in Berkshire and Oxfordshire, and indeed they were often associated with the various local amenity societies that sprang up. For example, Sir William Mount (Vice-Lieutenant of Berkshire and Vice-Chairman of the county council) was associated with the Kennet Valley Preservation Society; the Hon. Peter Remnant (MP for Wokingham 1950-9) and Colonel P. Fleming (President of the Oxfordshire branch of the Country Landowners' Association) with the Chilterns and South Oxfordshire Preservation Society; John Smith (Deputy Governor of the Royal Exchange Assurance, Director of Coutts and Co., Rolls-Royce Ltd, the *Financial Times*, and MP for the Cities of London and Westminster, 1965) with the Mid-Thames Association; Gerald Palmer (MP for Winchester 1935-45) and Lord Iliffe, the newspaper proprietor, with the Berkshire Downs Villages Association. Many other members of these societies were men of substance and standing – subscriptions to the Downs Preservation Committee, for example, were based upon acres and racehorses owned. In the circumstances, it was not unnaturally believed that there was a great deal of string pulling on the part of the 'landed gentry'. There is nothing to suggest that the Ministry of Transport was open to personal influence of this kind. It remains true, nevertheless, that men with experience of public life are likely to know which arguments will carry weight with government departments and how best to present their case.

10 Reading's reaction to this, of course, was that it was not so much a question of using the M4 to solve its own problems as of persuading the Minister not to fix the line in such a way as to *increase* the town's difficulties.

11 Oxfordshire, on reflection, had come to the conclusion that a route north of Reading would help solve its own traffic problems. It therefore changed sides and announced that it would support the Minister's proposals for a Revised Direct route. *Reading Mercury*, 18 July 1964.

12 H.C.Deb. (1963-4), 697, c. 1333-4. Furthermore it seems likely that by this time the Ministry of Transport had become very conscious of the problems created when motorway traffic is decanted at a single point on the rim of a conurbation: a number of feeder routes from south London to a second motorway would obviously help to ease the pressure that would otherwise have built up at the end of the M4 in west London.

13 It seems unlikely that a new Minister and a change of Government had any effect on departmental thinking; the reappraisal had begun long before Mr Marples left office. Nevertheless, when a proposal, the merits of which have been defended through thick and thin for a number of years, is to be thrown over, the advent of a new set of political chiefs makes the exercise a good deal easier if only because they personally have no words to eat. Probably of more importance was a change in personnel, at a very senior level, on the engineering side of the division within the Ministry of Transport that was dealing with the M4. This change occurred in April 1964, shortly before the consulting engineers' final report was received. The best recipe for solving old problems may well be to confront fresh minds with new evidence.

14 A draft scheme for the Maidenhead-Theale section was published in February 1966. Though they accepted a southern route in principle, Reading, Berkshire and the Cookham RDC all lodged formal objections to various detailed aspects of the scheme actually proposed by the Minister.

15 At Easter 1961 there was so much traffic on the A6 linking the motorways by-passing Preston and Lancaster that eventually traffic on the Lancaster by-pass was brought to a standstill, two deep, for three miles. See James Drake and R. A. Kidd, 'Planning for road traffic in the counties', *Planning for Traffic*, Institution of Municipal Engineers, 1961. After the opening of the Severn Bridge in September 1966 a great deal of through traffic from London to South Wales switched from the A40 to the M4/A4 route. Since the central section of the M4 was still uncompleted, this traffic, instead of by-passing Reading, began to pass through the town. Within a month congestion had worsened appreciably. Some lorries were taking an hour to drive through the town and the public transport services were in considerable difficulties.

16 The proposals made by the Ministry of Transport in August 1965 for the line of the M4 to the north of Newbury provide a striking illustration of this possibility. Although on a map they appear to be separate villages, Chieveley and Downend are in fact part of the same community, and in order to bind them together it was proposed to develop a recreation ground between the two. The consulting engineers, it appears, understandably mistook the recreation ground for an open field, and proposed to drive the motorway straight through it, thereby effectively cutting the village in half. After consultations with the local authorities involved the Ministry agreed to alter the line. *Newbury Weekly News*, 25 November 1965.

# Chapter 7

# The Juggernauts: public opposition to heavy lorries[1]

## Richard Kimber, J. J. Richardson and S. K. Brookes

> Frequently . . . a regular flow of heavy lorries is routed along roads [which] . . may not have been constructed to carry it. In the experience of a number of societies an observable result of this has been superficial or structural deterioration in the fabric of adjoining properties.
>
> Civic Trust report, *Heavy Lorries*, October 1970

> Property owners could, of course, help themselves by keeping their premises in a reasonable state of repair instead of waiting for them to fall down around their ears and then blaming lorries for it.
>
> 'Free wheels', *Roadway* (Journal of Road Haulage Association), January 1972

These statements reflect the intensity of the controversy over heavy lorries, and reveal the bitterness felt by some sections of the road transport industry at the end of the first stage of the long campaign to prevent the appearance on British roads of commercial vehicles in the 44- to 56-ton weight range. With the growth of environmental awareness, and concern regarding civic and social amenity, proposed changes in regulations which formerly would have gone unchallenged, such as those affecting transport and road use, are now the source of heated controversy and extensive public opposition. Here we propose to trace the growth and development of the opposition to increased lorry weights, and examine the political aspects of a controversy which, in the light of Britain's European aspirations, has now achieved international dimensions. In recent years the grudging toleration shown by the general public towards the heavy commercial vehicle has given way to vocal and in some cases highly emotional criticism. Now characterized as 'leviathans' or 'juggernauts', heavy lorries are held to be a source of danger, congestion, excessive noise and smoke, and as the cause of significant physical damage to the urban and rural environment. This change in public attitudes reflects the continuous growth in the number and size of commercial vehicles, and the attendant problem of their accommodation by a road network already under severe pressure from the ever growing volume of private cars.

Between 1956 and 1968 the number of goods vehicles with unladen weights of 5 tons or over (and therefore capable of 20- to 32-ton gross operating weights) increased from 34,000 to 150,000, and throughout this period the Ministry of Transport sanctioned a number of increases in their permitted dimensions and gross operating weights.[2] The incidence of complaints over nuisance, blight and damage has continued to grow despite motorway development and by-pass schemes, and regulation by the Ministry of the noise and smoke characteristics of heavy lorries.

A number of statutory and voluntary organizations have expressed the public concern regarding the depredations of the heavy lorry, and have criticized what they view as an 'unaccountable' road transport industry, reaping economic benefits at the expense of the community's environmental welfare. Leading conservationist organizations (the Council for the Preservation of Rural England [CPRE], the Society for the Protection of Ancient Buildings [SPAB]) and the voluntary associations concerned with road safety have pointed to the increased social costs caused by heavy lorries. These costs result from vibration, impact damage, accidents, road deaths and visual intrusion. The local authority associations have also been concerned by the escalation of road maintenance costs which must be borne, in part, by the ratepayer. On the other hand, the industry views these criticisms as 'latter-day Luddism', a reactionary attempt to thwart progress and further disrupt an economy which is dependent on cheap, and therefore heavy, transport. Comments such as those above, it has argued, are less accurately described as 'pro-environmental' than as irrationally 'anti-lorry'.

The problems associated with heavy commercial vehicles do not lend themselves to simplistic solutions such as 'keeping them off the road' or 'sending the traffic by rail', proposals which are the cause of much irritation amongst both civil servants and members of the road-transport industry. Equally clearly, the scale of the problem can only be met by intervention by the Central Government and the introduction of regulations which would modify not only vehicle noise and pollution characteristics but also the distribution system of which these vehicles are a part and the road networks to which they have access.

An opportunity to impress on Whitehall the need for regulations of this kind came in 1969 when proposals, originating from the vehicle manufacturers, raised the possibility of even larger and heavier commercial vehicles. For a number of years the organizations representing the various sections of the transport and road industry had been pressing the Government, on economic grounds, for increases in the permitted gross operating weights of their units. In July of 1969 the Society of Motor Manufacturers and Traders (SMMT) – following lengthy consultations with the technical sections of the Ministry of Transport – proposed to the Government that the Construction and Use Regulations governing the weights and dimensions of commercial vehicles should be substantially modified. These proposals came only one week after both the SMMT and the Ministry had described suggestions that the proposals were forthcoming, as premature, and initially were made with the other branches of the industry 'in the dark' as to their exact nature.[3] Following their publication, however, both the Road Haulage Association (RHA), representing the hauliers, and the Freight Transport Association (FTA), representing industrial and commercial organizations operating transport on their own account, automatically registered their support for the suggested amendments. The Construction and Use Regulations, in addition to stipulating axle weight limits and performance and braking standards, etc., specify for both rigid and articulated vehicles a scale of gross operating weights governed by a schedule involving stepped increases according to the number and spacing of axles. In 1964 when the existing limits came into operation, the maximum gross operating weights had been set at 28 tons for rigid vehicles with four axles and an outer axle spread (OAS) of 26 ft, and 32 tons for four- and five-axled articulated vehicles with spreads of 32 and 38 ft.

The SMMT proposals were for a full scale stepping of the permitted gross operating weights over the range of outer axle spreads and increasing the maximum limits to:

1 30 tons in the case of four-axled rigids with an OAS of 22 ft.
2 34 tons for four-axled articulated vehicles with an OAS of 31 ft.
3 44 tons in the case of five-axled articulated vehicles with an OAS of 40 ft.

The SMMT also recommended that rigid vehicles and drawbar trailer units (lorry-trains) be permitted to gross 56 tons in contrast to the 32-ton limit which currently applied. Although the Society made special note of the fact that it was making no appeal for any increase in the existing 10-ton axle limit, it did propose that the tandem axle distance for 20 tons, which stood at 7 ft, should be reduced to 6 ft 3 ins and that the weight limit for tri-axles be raised to 24 tons.[4]

There was nothing to prevent the manufacturers from producing, plating and marketing models capable of grossing the increased weights, as their dimensions, noise and smoke characteristics still conformed to the existing standards laid down by the Construction and Use Regulations. In fact their production had already begun and the leading manufacturers – Guy, Scammell, AEC, Atkinson and ERF – exhibited their '44 tonners' at the Scottish show only three months after submitting their proposals to the Ministry of Transport.

British hauliers were, even so, unwilling to invest their capital without Government approval of the higher weight limits. To some degree the operating costs of the new vehicles were higher than normal and if circumstances went against them they risked being stranded, holding units which could only be operated at below maximum efficiency.

Essentially, however, the 44 tonners were aimed at the Continental market, because in the Soviet Bloc, the EEC, Scandinavia, Spain, Portugal and Austria gross operating weights of 38 tons or above were permitted. This discrepancy formed the basis of the SMMT's case which it submitted to the Ministry in support of its proposals. In the memorandum in which the SMMT summarized its case,[5] it argued that 'successive Governments had urged the harmonization of international regulations affecting motor vehicle design', and that the suggested modifications would bring the UK closer into line with current general practice in Europe. Most importantly, it stated that if the proposals were rejected 'vehicle manufacturers would continue to lose valuable commercial opportunities in world markets'.[6] It was, the SMMT argued, impossible to develop and market the higher payload vehicles, 'required for sale in overseas markets at competitive prices, without the British home market supplying the volume base'.[7] Together with the haulage associations, the SMMT pointed to the Government's commitment to the policy of containerization and the necessity for higher gross weights if 20-ft, 30-ft and 40-ft containers were to be fully and economically utilized. As far as the haulage industry was concerned, the proposed modifications held forth the prospect of general increases in carrying capacity and productivity and hence greater returns on the capital and labour it employed. It was also argued that savings in cost per unit transported could be passed on to the consumer, or at least enable spiralling freight costs to be contained.[8]

The road-transport industry was confident that its recommendations would be accepted and then presented to Parliament before the end of the year. As early as September 1968, the then Minister of Transport, Richard Marsh, writing in *Motor Transport*, had indicated his awareness 'that many operators would like the Government to permit greater maximum weights . . . to keep pace with developments like containerization'. He was, he declared, 'ready to give the manufacturers the go ahead on this kind of thing providing they were prepared to play their part in minimizing the undesirable social side effects'. In January 1969, a Ministry of Transport spokesman had said that progress had been made in reviewing higher gross weights, bearing in mind the effects these would have on roads and bridges. He hoped it would not be long before definite proposals could be announced.

After the industry had first indicated its desire for weight increases, and following the preliminary results of a roads' and bridges' survey, serious talks between the Ministry and the manufacturers began. The results had shown that objections on the grounds of road capability would be unfounded. In effect, Ministry officials had guided members of the technical committee of the SMMT in their formulation of the proposals and agreement had already been reached with regard to the technical aspects of the Society's recommendations.[9] There was much, therefore, to confirm the optimistic expectation of the manufacturers that an early and favourable decision would be reached.

With one significant exception, publicity for and discussion of these proposals were confined to the trade press. This exception was the *Sunday Times* which, after reporting the manufacturers' representations, devoted regular space (for over one-and-a-half years) to a series by Tony Dawe, bearing the title: 'The Juggernauts.' The general tone of these articles was that of opposition to the road-transport industry. Priority was given to statements made by objectors and the case against the increases on amenity and environmental grounds was presented. At this stage Dawe shared the manufacturers' belief that the Government would accede to the industry's requests at a time when it was implementing a Transport Bill with its emphasis on greater productivity. He wrote that 'Richard Marsh is certain to increase the limits to a standard continental figure of 36 tons.'[10]

Notwithstanding this fatalism, the continuing interest of the *Sunday Times* was instrumental in transforming the unobtrusive and unpublicized proposals of the road-transport industry into a controversy of national significance. Throughout the following months, its extensive coverage contributed to the growth of public awareness and the creation of an atmosphere in which it was impossible for the Transport Ministry to arrive at a decision without careful consideration of factors other than those of technical feasibility and economic advantage. In giving prominence to this issue, the *Sunday Times* effectively reinforced and then mobilized the latent opposition towards heavy lorries which existed amongst its middle-class readership.

An additional effect of the prominence given to the 'heavy lorries' problem by the *Sunday Times* was to maintain interest throughout the critical period of the Parliamentary summer recess. When in October 1969 Parliament reassembled, several Members were ready with pertinent questions to put to the new Minister of Transport. By this stage most of the other national newspapers felt the issue to be worthy of space, and the regular coverage which followed thereafter expressed (with the exception of *The Times* and the *Financial Times*) general disapproval of the proposed increases.

The early newspaper interest also served to give publicity to the initial reactions of the opposing voluntary organizations. The National Council on Inland Transport (a study group which had for many years advanced arguments for the economic and social benefits to be gained from a reorganization of inland freight transport) pointed out that, on the basis of a survey undertaken by US Highway officials, the proposed increases would treble the damage caused to road surfaces. The Pedestrians' Association for Road Safety, whose interest in matters of road transport is broader than its name would suggest, stated that it was time to call a halt to such increases and tell the transport industry to split up its loads or send them by rail. John Connell, Chairman of the Noise Abatement Society, claimed that lorries had caused more damage to ancient buildings in the last fifty years than had been done in the previous 1,000. This, he claimed, was due to noise and vibration and he could not understand why the Government was considering such proposals when the roads were not good enough.

These were the first public attempts to widen the terms of the argument over the proposed weight increases – then confined to considerations of a technical nature – and thus forestall an early and unfavourable decision by the Minister of Transport. In this respect, however, greater success had been achieved by a direct approach to the Minister himself, made on behalf of one of the provincial civic societies. Many of these societies shared a common concern over the growth in the numbers of heavy lorries, but in particular the Faversham Society decided that more positive action was needed. It was fortunate to have as its Secretary Arthur Percival, the Civic Trust official whose special responsibilities were road use and transport problems, and who was immediately alive, therefore, to the implications of a bid for further weight increases. Noting that neither the Civic Trust nor the CPRE had ever been consulted regarding amendments to the Construction and Use Regulations, the Society approached Terry Boston, then MP for Faversham, who took the matter up with the Minister of Transport and secured an agreement that the two organizations would be consulted regarding future proposed changes.[11]

Despite this commitment, public protest continued to increase. For example further representations were made by the Pedestrians' Association and by the Joint Committee of the SPAB, the Civic Trust, the Georgian Group and the Victorian Society.[12] The Georgian Group, after considering a resolution from the Colchester Civic Society, had also asked the Ministry for consultation rights.[13] As a result of these developments, the Minister began to modify his position towards the industry. At a press conference, soon after his appointment as Minister of Transport, Fred Mulley indicated that an early decision on the weights issue would not be forthcoming as it would take a considerable time to consult all the parties concerned. This was the first admission that the Ministry felt that it was no longer able to give unqualified approval to the SMMT's proposals.

On the day following a letter to *The Times* 'voicing the growing public alarm' at the proposed weight increases (signed by leading members of the amenity societies in both the Commons and the House of Lords[14]) the new position was reaffirmed by the Minister. Answering a question from John Ellis (Bristol, North-west), Mulley adhered to the view that the proposals had originated not from the Ministry but from the manufacturers. He said that the Department was looking into the safety aspects and was discussing the technical implications with the industry, and implied that his deliberations were at an early stage. He gave an assurance that 'as soon as there was something positive to discuss he would consult with all other interests including those concerned with safety and amenity'.

If these reassurances were intended to disarm the growing opposition, they were singularly unsuccessful. In effect, these ministerial statements served only to give greater publicity to the controversy and to heighten the awareness and intensify the efforts of those opposed to the industry's proposals.

The CPRE – which for a long time had been concerned over roadside and surface damage and the continual loss of rural land to motorway and road widening schemes – opened what was to be a substantial press campaign with a letter from its Chairman, Lord Molson, aligning the CPRE with the views earlier expressed by the

Bishop of Chester. The CPRE, in a lengthy press statement following this letter (which received extensive national coverage) took the realistic view that containers were the obvious solution to transport in the future but argued that if 44 tonners were to be used they would have to be confined to motorways and other main arterial roads. The statement concluded, however, with the more emotional charge that 'if lorry loads were to be further increased, then life would become intolerable'.

More direct efforts were made to mobilize public opposition by the Pedestrians' Association,[15] which in November convened a public meeting in Westminster on the subject of the lorry and the community. This meeting was chaired by the Earl of Kinnoull and addressed by Tom Foley (himself a Vice-President of the NCIT), A. F. Holford-Walker of the CPRE, Leslie Lewis of the Society for the Protection of Ancient Buildings, John Connell, and Peter Nottley of the Cyclists' Touring Club. The resolution passed by the meeting 'opposed any increases to the existing weight and length of goods vehicles which would inevitably pose a direct threat to a civilized environment in both town and country'. At this meeting John Connell argued that democratic government 'goes in the direction in which it is pushed by the greatest force from whichever source it comes. Unfortunately, the economic source is by far the greatest and the motor-vehicle manufacturers said that what they wanted was economically necessary. The Ministry of Transport, not knowing any better, said that if that were so then they would agree.'[16]

On the issue of heavy lorries, however, the pressure which the amenity bodies could exert was substantially reinforced by that of several influential organizations whose opposition to the proposed weight increases was based on economic considerations which somewhat conflicted with those of the motor-transport industry. These were the representative bodies of the various branches of local government who, as highway authorities, had as one of their chief concerns the escalating road repair bills resulting from the increase in heavy-lorry traffic. As regular and responsible consultees, their views carried a certain amount of weight with government departments and the necessary steps were taken to make these known. Noting Mulley's parliamentary statement, several member authorities of the Association of Municipal Corporations expressed their concern over the possibility of even heavier lorries using unsuitable highways and representations were made to the Ministry by the Association's Secretary. The County Councils, as highway authorities for the non-urban areas, were also conscious of the costs of damage to country roads, and their opinions were communicated to the Ministry by the County Councils' Association. In November

1969 a resolution came before the National Association of Parish Councils, at the National Conference of Parish Councillors, expressing the hope 'that the Ministry of Transport would bring into force legislation prohibiting the use of any vehicle larger or heavier than those presently permitted'. Subsequently, this resolution was formally adopted by the Council of the Association (together with a call for the NAPC to support the campaign of the CPRE) and submitted to the Government. The Urban and Rural District Councils' Associations also stated its objections on the grounds of cost, damage to underground public services and insufficient knowledge about the effects of noise and vibration. They further advocated the creation of a designated routeing system to channel traffic away from the damage-prone areas.

Whilst, for the most part, firmly denying that they took part in a campaign against the Government or the transport industries, the local authority associations contributed greatly to the mounting public and private pressure with which the Government was faced. Indicative of the opposition, was the degree of parliamentary interest in the weights proposals. MPs had been receiving letters from their constituents expressing their fear of further licence being given to the transport industry, and now began to voice these fears in Parliament. In November, MPs of both parties spoke predicting 'the widespread dismay there would be if some of the proposals being suggested by the road hauliers are accepted'. Joan Quennel, MP for Petersfield, gave her personal support to the residents of Droxford (Hampshire) in their presentation to the Minister of Transport of a memorandum on the town's traffic problems, and other MPs passed on their correspondence to convince the Ministry of the political undesirability of a decision favourable to the transport industry. In December, both Houses debated the proposals; the opposition in the Lords being led by Lords Moyle, Canesford and Popplewell. In the Commons the debate was opened by Graham Page, and Sir Gerald Nabarro proposed that 'if vehicles of the 44 to 56 tons range were to be considered, they ought to be confined to roads of a motorway type'. The Liberal MP, Peter Bessell, noting the increase in the number of accidents involving heavy lorries, asked Fred Mulley outright 'for an undertaking that he would resist any proposals for an increase in tonnage or length'. This request was greeted by cheers. The line taken by Fred Mulley and the Government spokesman in the Lords – Lord Shepherd – was that the proposals had been made by responsible organizations who played an all-important part in Britain's ability to make a living and whose views required careful consideration. Fred Mulley did acknowledge, however, that he had received objections from a number of organizations and renewed his promise to consult the many interests involved before authorizing any change in the regulations.[17]

This promise was fulfilled in January of 1970 when the Ministry issued an invitation to the objecting organizations to attend a meeting on 10 February. This was 'to afford an opportunity for organizations outside the road transport industry to express their views on the proposals made by the industry'. The invitation note stressed the Ministry's desire to take into account the views of all concerned and its willingness to receive written evidence in support of any opinions expressed at the meeting.

Invitations were addressed to all the representative bodies, whose opposition we have already noted, and to a number of others who had also shown interest in this topic either currently or on previous occasions – the Royal Institute of British Architects (RIBA), the motoring organizations, the Countryside Commission, the Ramblers' Association and the Royal Society for the Prevention of Accidents (RoSPA). The meeting was convened as a direct response to the volume of pressure being felt by the Ministry. A Ministry official has confirmed that several hundred individuals wrote to the Department or their MPs on the lorries issue, in addition to the official representations of the amenity bodies. Besides the opposition in the Commons and the pro-environmental line taken in the national newspapers, the departmental solicitude for the industry had been further undermined by the publication of a report by its own agency, the Road Research Laboratory, stating that heavy lorries were responsible for a large part of the £100m. per year road maintenance bill and for delays to other road users costing an estimated £10m. per year.[18]

The Ministry's decision to hold this meeting in private was considered by John Connell to be an 'arrogant and patent attempt to stifle comment'. But it is clear from notes of the meeting that if this was the desired effect it was not achieved. The organizations presented a broad front of opposition (with variations in intensity in the conditions that they felt must obtain before increases were contemplated). Three-quarters of the delegates proposed designated or restricted route systems, excluding such vehicles from residential streets, town centres, conservation areas and minor rural villages. The local authority associations stated their objections at length and the RIBA suggested that the industry's proposals be subjected to a thorough cost benefit analysis.[19]

A similar meeting was held in Edinburgh on 6 March, attended by the equivalent Scottish organizations (e.g., the Association for the Preservation of Rural Scotland, the Scottish Civic Trust, the County Councils' Association in Scotland, the Convention of Royal Burghs). Opinions in Scotland were more diverse. Unqualified opposition came from the Scottish Counties and Cities Association and the Scottish Railway's Development Association. The degree of English interest was not shared, however, and the Association of County Councils in Scotland, the National Trust for Scotland and the Association for the Preservation of Rural Scotland stated that they had not had an opportunity to examine the industry's proposals in detail. At the other end of the spectrum, the Highlands and Islands Development Board wholeheartedly supported a new scale of weights, believing that they would assist in the economic development of the Highlands.[20]

The Scottish Civic Trust was merely concerned that close attention be paid to the design of the proposed vehicles to ensure that they were more manoeuvrable, thereby reducing congestion and impact damage in city centres. This was in marked contrast to the deep concern shown by its English counterpart. At the 10 February meeting, the Civic Trust had objected not only to increased vehicle weights but also to those presently permitted. Prior to the meeting the Trust had consulted the 700 local civic societies for their views on the proposals and for information on the environmental disruption caused by the heavy lorries already on the roads. Over 300 of these societies submitted detailed views and evidence to the Trust and on a brief analysis of the returns, a preliminary memorandum was prepared for presentation at the meeting. Arthur Percival, who had been in charge of this operation, then set about undertaking a more detailed analysis of the evidence and preparation of a fuller memorandum for submission to the Ministry.

When this report was finally presented in September 1970, it contained the most comprehensive catalogue of lorry nuisance and damage that had hitherto been compiled and provided the Ministry with a ready answer to the 'cogent economic arguments' of the transport industry.

Although evidence for the report was also supplied by the Joint Committee and the CPRE (from Headquarters and from its forty registered branches), which had earlier launched its own nationwide investigation into heavy-lorry damage, the most significant feature of the exercise was the appeal to the local civic societies. This marked a change from the usual procedures employed by the Trust. Generally, the Trust has confined its attention to issues of topicality and to giving publicity to those whose achievement in conservation and civic beauty could be emulated. The mounting of an opposition campaign and the mobilization of local groups is a departure which could be repeated to the advantage of those with amenity and conservation interests at heart. The success of the operation had depended on the returns from the provincial societies and the 45 per cent response achieved in the eighteen-day period before the meeting with Ministry officials attested to the degree of concern and efficiency of these organizations. It would appear also, from the failure of the local authority associations to present any detailed memoranda to the Ministry, that the civic societies were far more capable of collecting the necessary evidence in the short time available than the local councils. In undertaking such work, the official bodies lack the energy and manoeuvrability of voluntary societies and are hamstrung by their procedures, limitations of time, and by the restrictions placed on the initiative of elected and administrative officials.

In the preparation of its report, the Trust pursued a policy of almost deliberate isolation and one of the most interesting features of the opposition's campaign was the lack of co-ordination on the part of the numerous organizations concerned. Even so, efforts were made to present a co-ordinated viewpoint. The standing committee of the major amenity organizations, CoEnCo (which includes the Civic Trust, the CPRE, the Noise Abatement Society, the Ramblers' Association and the SPAB) set up a working party to 'co-ordinate the presentation of the amenity case to the Government'.[21] In the event, this committee undertook no work of importance and adopted the CPRE report on heavy lorries for the CoEnCo 1970 Annual Report. Its work had in effect been pre-empted by the report of the Civic Trust, which was already underway. The official explanation for the committee's inactivity was that such committees were often set up as watchdogs, in case constituent members were not able to deal with a particular issue adequately. As far as the 'heavy-lorries' problem was concerned, this was not the case, and therefore the committee was allowed to lapse.

Other bodies followed up their participation at the February meeting by submitting independent memoranda. The Pedestrians' Association outlined its objections and emphasized the growing volume of protest from all parts of the country against the danger and nuisance created by heavy-lorry traffic. It maintained its opposition by forwarding to the Minister, in July, a copy of an article on heavy lorries by the Association's Chairman. The RIBA elaborated on its demands for a cost benefit analysis of any increases in lorry weight and size – the onus to prove the necessity for change, it claimed, was unequivocally laid on those seeking to raise the limits.[22] The Institute registered its opposition on the grounds of corresponding increases in the physical damage to buildings by collision and vibration, the inability of new models to operate within tolerable noise and pollution limits, and the inadequacy of the existing road network.

The RIBA memorandum revived the (by then) flagging interest in the national press, and was the theme of several articles generally opposed to the industry's proposals. Only the *Financial Times* gave prominence to the reply put out by the SMMT. An interesting feature of the long controversy was the road-transport industry's inability to allay popular disquiet. The general line taken by spokesmen for the SMMT and RHA was that the widespread talk of 'bigger' lorries was misleading. As far as the layman was concerned the new vehicles would be indistinguishable from those operating at present – and in some cases were already operating, but below maximum efficiency. Emphasis was placed on the fact that the proposals did not involve an increase in maximum axle loadings; in fact, the axle loading of existing vehicles was often exceeded. There was also a possibility that higher gross weights would lead to a reduction of the number of vehicles on the road. The new lorries, it was argued, would have improved braking efficiency and safety standards and would comply with the existing regulations regarding noise and smoke emission and there was no reason to believe they would prove of any further detriment to the environment. They had got to come in the interests of progress and efficiency.

The transport industry's position was, of course, a difficult one. As we have pointed out, given the general unpopularity of the lorry, it was impossible to mount any effective public-relations campaign which could transform its image and generate support for the proposals. Traditionally the Conservative Party has been closely associated with transport interests and on previous occasions the industry has not lacked parliamentary support. On this occasion, however, few of its friends were willing to come to its assistance on an issue which provoked such intense feeling at constituency level. Also, the failure of the road-transport industry to secure parliamentary defence of its proposals may be attributed to the dependence of the SMMT on contacts with government departments. As William Plowden has pointed out, the road-transport industry and the SMMT especially 'is engaged in a continuous process of formal and informal consultation which makes much Parliamentary activity superfluous'.[23] Efforts in the parliamentary arena had also not proved rewarding. One arm of the industry had attempted to cultivate MPs by entertaining them – the end result had been a massive drinks bill and little in the way of parliamentary support.

Several further RRL reports were the source of further ammunition for the environment and amenity lobby. In May 1970, a report analysing fatal accidents involving cars and light vans found that in 50 per cent of the cases the other vehicle involved was a heavy commercial lorry.[24] These findings confirmed the results of earlier studies which had pointed to the disproportionate number of fatalities resulting from heavy-lorry accidents (i.e. in relation to the total number of accidents and proportion of heavy goods vehicles in the vehicle population).

In November 1970 a report of a RRL working party – 'A review of road traffic noise'[25] – recommended that 'for maximum noise reduction, heavy lorries should be confined to main roads except where local access was required'. In its Annual Report for 1969, the Laboratory also referred to a study carried out on motor traffic and stated that although only 1·7 per cent of the total number of axles carried loads of more than 10·7 tons, they contributed about 30 per cent of the 'damaging power' of the traffic. It stated that 'analysis of effects of possible changes in permissible weights of commercial vehicles was hampered by a lack of reliable statistics on operational costs and an absence of adequate knowledge as to how much costs are affected by changes in vehicle characteristics'.[26]

A further blow for the amenity organizations was struck in November 1970 with the publication of the Marshall Committee Report on Highway Maintenance. This stated that the use of weak and unsuitable roads by heavy lorries added considerably to the road maintenance bill and recommended that the appropriate Ministry look into the possibility of national legislation to confine heavy vehicles to major roads. It also recommended that local highway authorities should explore the possibility of traffic regulation orders. It is difficult to measure the impact that these reports had on departmental thinking but, given that a decision was eventually reached without the benefit of the specially commissioned cost benefit report from the RRL, it would appear that such documents were of limited significance.

Their importance lay not so much in being a basis for policy-making but in providing further evidence for the amenity lobby's attack on the transport industry and the Department, and adding to the weight of the political considerations that the responsible Minister had to take into account.[27]

Before the decision was taken, however, the location of responsibility changed in several ways. The Conservative General Election victory resulted in the replacement of Fred Mulley by John Peyton (who, it could be discerned later, had some degree of sympathy for the road-transport industry). Of even more importance were the departmental changes following the lines of the new Government's White Paper *The Reorganisation of Central Government*.[28]

The merging of the Ministries of Housing and Local Government, Transport, and Public Buildings and Works, establishing a department responsible for the whole range of functions affecting the 'living environment',[29] placed the final responsibility for a decision on the weights issue in the hands of the Secretary for the Environment, Peter Walker.

Paradoxically, for conservationists in general, as well as those solely concerned with the weights issue, the chances of success seemed to have been drastically reduced. Conservation and amenity group leaders had felt at home with the Labour Government and regarded Labour Ministers as amenable as far as environmental problems were concerned. The identity of interest between the Conservatives and industry (and the road-haulage industry especially) seemed to bode ill for the future. This dismay was compounded by the appointment of Peter Walker, himself a successful businessman, who during his years in Parliament had failed to indicate any concern regarding problems of environment and amenity and had in 1967-8 supported the industry as opposition spokesman on transport during the debates on the Transport Bill.

The industry renewed its efforts to obtain an early decision from the new Ministers but was informed that the RRL tests were running a month behind schedule and that there was no hope of any official announcement on the subject of higher weights before October. Representations were made by the technical committee of the RHA, urging the Government to introduce regulations based on the new axle spacing theories for vehicles of up to 32 tons. These, the RHA argued, could be introduced before waiting for the domestic and international political arguments regarding vehicles over 32 tons to be settled. These points were put directly to the Minister when William McMillan (Chairman) and G. K. Newman (Director General) of the RHA met Peyton for preliminary talks when the new incumbent was 'getting to know the major organizations with which he would have to deal'.[30]

The SMMT also renewed its efforts and, with some appreciation of the problems faced by the Ministry, had now come round to the view that the best solution would be a 'package' whereby a new schedule of gross operating weights would come 'wrapped' in further regulations dealing with related environmental factors – noise, smoke and minimum power to weight ratios, etc. With this type of agreement in mind, the Minister for Transport Industries informed the SMMT that if they would reduce some of the pressure on himself and his Department the 'door would be kept open' for new regulations of the type for which they had been pressing. The manufacturers took this as a commitment – albeit informal – binding the Minister to introduce a new schedule on the basis of their initial proposals.

With rumours, originating from the transport industry, suggesting that a decision on the weights issue was imminent, press coverage of the controversy revived in September 1970. This coincided with the publication of the CoEnCo report. The report, which was passed on to the Ministry, concluded that there should be no further increases in vehicle load carrying capacity in view of the damage already being caused by vehicles operating within the existing weight limits.[31] Introducing the report, Peter Robshaw reiterated another of the points emphasized by the CPRE and the other amenity organizations. This was the tremendous volume of concern that the threat of higher weights had provoked in both urban and rural areas.[32] It had become clear that a decision favouring the road-transport industry would fly in the face of a substantial volume of public opinion. For the conservation movement as a whole, the issue had become one of major significance. In the light of the disillusionment over the Samlesbury Brewery decision (Peter Walker, as Minister, had given Whitbread permission to build a brewery in open country at Samlesbury, Lancashire) the heavy-lorries question had now, in Professor Kenneth Denbigh's words, 'become the acid test' of the Conservative Government's intentions regarding problems of conservation and environmental welfare.[33]

The Ministry was, however, taking steps to emphasize its concern and indicated that its decision would not be taken without the fullest consideration of the environmentalist case. Having accorded consultation rights to the major organizations, the Ministry began to press them to look to their responsibilities and several times requested the Civic Trust to provide the detailed evidence which it had promised. The Trust eventually submitted its comprehensive report 'Heavy lorries' to the Ministry in late September. The report said little that was not already part of the opposition case, but constituted the broadest and most analytical summary of heavy-lorry damage, supporting its claims with a wealth of written and photographic evidence.

The Department had already considered most of the points that it raised, as they had either been put forward at the two meetings, or had been discussed internally. Again, as with the RRL reports, its significance lay in its usefulness (when eventually made public) as a 'stick' with which to beat the transport industry and as a further indication – having been compiled from provincial civic society returns – of the widespread public opposition to weight increases.

Further efforts were made to obtain the maximum amount of publicity for the opposition case, and to place the Government under the greatest degree of pressure. CoEnCo announced that it was to meet representatives of the Pedestrians' Association and would attempt to raise the issue again in Parliament. The CPRE also reopened its press campaign and, speaking on the first day of the third 'Countryside in 1970' conference (attended by senior members of the Government), Holford-Walker (Joint Secretary of CoEnCo and the CPRE) called for 'an absolute ban on any increase in the load bearing capacity of heavy lorries'.[34]

In early November, Lord Mowbray, the Government spokesman on transport in the Lords, indicated that a decision would be made in the near future. This produced a month of fierce speculation with regard to the Government's intentions. Claiming that Ministers had now accepted the economic arguments for weight increases, the *Sun* headlined an article 'Road monsters to get go ahead'. The *Mirror*, on the other hand, confidently predicted 'Britain's no to monster lorries'. As Tony Aldous has noted, however, in retrospect the clearest indication of the nature of the Government's decision came on 6 December 1970 at the press conference called by the Civic Trust to announce the publication of its 'Heavy lorries' report. At this meeting the Trust's Chairman, Duncan Sandys, MP (himself a former Conservative Cabinet Minister with many years of governmental experience, and still a major figure in the ranks of the Conservative Party), spoke with optimism regarding the outcome of the weights issue. His confidence that 'good sense would prevail' and that no further public campaign was necessary to convince Ministers, is likely to have been the result of prior reassurances that a favourable decision was forthcoming.[35]

In the national press, virtually unanimous editorial opposition to the weights proposals followed the publication of the Civic Trust report. At Westminster on 8 December, the Pedestrians' Association convened a meeting between an *ad hoc* committee of objecting organizations and interested MPs. On the parliamentary side this meeting was sponsored by an all-party group comprising Angus Maude (Cons., Stratford), John Pardoe (Lib., Cornwall) and Arthur Blenkinsop (Lab., South Shields). In effect the efforts in and out of Parliament merely constituted a victory flourish. The SMMT had already been made aware that the Government had decided in favour of the amenity lobby. On 10 December, in a letter to John Peyton, the Society graciously conceded its defeat. It accepted that the Government could not agree to an increase in the maximum operating weights, for environmental reasons, but asked that prompt action be taken to 'introduce the agreed amendments in relation to the maximum gross weights and axle spacings for vehicles of up to 32 tons'.

The letter expressed the SMMT's willingness to recognize the political problems faced by the Minister, and hoped this would be matched by a recognition of the practical problems faced by the industry regarding the improvement of the noise and pollution characteristics of its vehicles.

On 16 December the Minister in reply to a question by John Rankin, announced his decision not to allow any increase in the maximum weights of goods vehicles. He was 'concerned that greater efforts be made to reduce noise, congestion, and pollution, and to this end he was circulating draft regulations on noise, smoke, and engine power which would come into effect in 1972/3'. The one concession made to the industry was a promise to draft regulations permitting marginal weight increases for certain specified vehicles within the 32-ton maximum. In the statement issued by the Department, Peyton thanked the industry for its helpful and co-operative attitude but stated that having given careful consideration to their views, he had concluded that the time had come to curb a growing and undoubted nuisance.[36]

In reality, the leaders of the road-transport industry were bitter in their disappointment. The informal agreement reached between the SMMT representatives and the Minister seemed to have been broken and the door, far from being left open, had been abruptly slammed in their face. The responsibility for this decision, however, lay not with the Minister for Transport Industries but with the Secretary for the Environment himself, who had intervened vetoing any package formulation drawn up by the Ministry.[37] The Secretary for the Environment had conceded that the weights decision had become the 'acid test' of the Government's environmental intentions, and that it was awaited by a growing and increasingly anxious environmentalist constituency. There were therefore substantial political benefits to be gained from bowing to public opinion at this time, and little or nothing to be lost from disappointing the manufacturers. It is perhaps worth noting that the Secretary's acute awareness of the extent of the environmental lobby had, in the period since the height of the heavy lorries controversy, produced a series of pro-environmental decisions for which the lobby had been pressing. In the eyes of his new constituency Peter Walker had, for the moment, re-established himself as a Minister with a helpful and imaginative contribution to make to the conservation movement.

The road-transport industry, on the other hand, viewed the Secretary in a somewhat different light. After attempting unsuccessfully to arrange a meeting to present their case, the industry's representatives gave up their attempts to secure a 'package agreement'. Publicly they expressed the view that the decision was only a temporary setback to an inevitable change towards larger and more economic transport units. The decision, they argued, could easily be reversed when Britain joined the Common Market. Privately they may have concluded, that as it was impossible to gain useful access to the Secretary, nothing of importance could be achieved until either he had left the Government or had been transferred to another Ministry.

A belief common to many members of the road-transport industry is that, from the industry's point of view, 'everything would be all right if it was simply a question of working and negotiating with permanent officials of the Ministry'. They, it is believed, are far more responsive to the industry's problem than their elected superiors. This situation may, however, not be permanent. As a result of their representations to the Ministry, the Civic Trust, the Joint Committee and the CPRE have now been accorded consultation rights and since 1970 have been consulted on all matters relating to transport and road use being considered by the Department. The significance of this development lies in the incremental effect that regular formal and informal contacts with representatives of amenity bodies will have on thinking within the Department. It may well result in a less sympathetic response to proposals originating from the industry.

The public outcry over heavy-lorry damage has already prompted the Government to take further steps to improve the situation. In the first months of 1972 the Road Traffic (Foreign Vehicles) Bill was presented to Parliament. Its provisions allow customs officials and police to detain Continental vehicles until they have complied with British weight regulations. At present, according to checks undertaken by customs officials and county highway authorities, a large proportion of foreign vehicles entering Britain (between 40 and 60 per cent) are substantially overloaded. It has been claimed that they are responsible for much of the worst damage, and for a large number of the complaints from the civic and amenity societies. Support for this legislation came not only from the environment lobby, but also from the haulage industry,[38] which was jealous of the freedom given to Continental operators and was anxious to focus public indignation on a foreign menace, thereby relieving pressure on themselves.

When the heavy-lorries problem had become a nationally important issue, the Department of the Environment had set up a Working Party on Lorry Parking to examine the problems of nuisance and intrusion resulting from the parking of lorries on public highways and in residential areas. Its report recommended the establishment by local authorities of a network of lorry parks, offering security and accommodation facilities and which could also be used as interchange points for the transfer of goods from large lorries to smaller vans more suitable for local collection and delivery. At the end of 1971 John Peyton gave official support to these proposals and announced that the Government would use its own powers (and not rely solely on those of the local authorities) to acquire suitable sites, and would make £10m. available for this purpose.

As we have noted, when announcing his decision not to allow further gross weight increases, the Minister had promised to circulate regulations allowing reductions in the axle spacings required for loads of up to 32 tons. These, however, were not released by the Ministry for over a year. According to the Assistant Chief Engineer at the Department of the Environment 'what should have been a relatively straightforward exercise proved far more complicated than expected mainly because of the safeguards required for bridges and roads'.[39] As it has been already noted, prior to the submission by the SMMT of its original proposals, it had already been decided that there were no technical problems regarding road and bridge capability. It would appear, therefore, that an examination was called for mainly to allay the fears of the highway authorities. The manufacturers held that this delay was responsible for the prevailing slump in the truck industry. Operators, they argued, were holding back orders until the Government had reached a decision. The circulation of these regulations, which allowed for a 2-ton increase in the gross weight of six- and eight-wheeled rigid vehicles, did produce full order books for some firms. The real sources of the industry's depression were, however, far more fundamental than uncertainty regarding weight regulations and for the industry generally the new weights were not the key to economic revival.[40]

Generally, the industry has recognized its failure to respond to the criticisms of the environmental lobby, but has concluded that there is little which can be done to improve the situation in a dramatic way. The FTA considers its best strategy to be the avoidance of gimmickry and concentration on commending to its members adherence to environmental regulations and a greater awareness of environmental problems. It is, on its own behalf and in conjunction with the Ministry, undertaking research into designated routeing systems and its Urban Traffic working group is presently studying the feasibility of night deliveries and standards for off-street loading facilities in urban areas. The industry still faces the problem of obtaining less critical press coverage. There are few clear statements of the industry's case outside the trade press but one notable exception has been the opportunity afforded by *New Society* for an academic to present those arguments (in favour of heavier lorries) which 'have not been given a proper airing'.[41] Essentially, however, there has been little change in press attitudes regarding the heavy lorry.

The Government, for its part, has adhered to the line established in December 1970. On several occasions throughout the following year John Peyton reiterated the points he had made in Parliament.

At luncheon addresses to both the RHA and the FTA the Minister explained that his decision was justified 'by the tide of public opinion that was beginning to favour the imposition of controls over the things which make life hideous'. 'It would be a mistake', he told the RHA, to regard such pressures as emanating from a noisy minority. He afforded the industry some consolation by holding forth the possibility of weight relaxations in the event of improvements in the motorway network and the routes to the major ports. These promises, when repeated in Parliament, met with strong criticism and press comment.

Developments at the international level had, however, undermined both the victory of the amenity societies and the Government's resolve on their behalf. The tide had turned strongly in favour of the road-transport industry, and had somewhat reduced the significance of the Minister's promise of concessions in the event of road improvements.

During the final stages of the controversy, the Conservative Government reaffirmed and strengthened its commitment to Common Market entry. For the transport industry part of the significance of this lay in the Community's efforts to honour the clause in the Treaty of Rome, which called for a common transport policy. The first steps towards this were the considerations, by the EEC Commission, of regulations to harmonize the widely varying gross operating and axle weights which obtain throughout the Community and which were considered a barrier to the free flow of goods. The Common Market norm stood at 38 tons (Dutch and Italian vehicles being allowed to gross 50 and 44 tons respectively whilst those of Luxembourg were limited to 36). Any compromise agreement which the EEC Commission was likely to present to the Council of Ministers would therefore be for a gross operating weight well above the British maximum.

Draft regulations were presented to the Council of Ministers in June 1971 recommending a gross operating weight of 42 tons and an axle weight of 11·5 tons. The British Government, therefore, in the light of its proposed EEC entry, was forced to give consideration to these proposals. As can be imagined, the road-transport industry was somewhat amused that 'less than a year after rejecting an increase in weights, the Department of the Environment was compelled to consult [the industry] on the desirability of an increase'.

The Council of Ministers considered the Commission's proposals in October and December 1971, but no agreement could be reached, mainly because of Germany's intractability on the question of axle weights. France, Belgium and Luxembourg, which permit a 13-ton axle limit, had agreed to the compromise figure of 11·5 tons. The

German Transport Minister had, however, announced 'that he could consider 11 tons but not a milligram more'.[42]

This failure to agree offered Britain a further opportunity to voice to the EEC her total opposition to any increase above the 32-ton gross weight or the present 10-ton British axle limit. It has been estimated that the cost of adjusting Britain's road network to accommodate the weights proposed by the Commission would be approximately £300m. together with a further expenditure of £10m. a year, on new roads – in addition to any additional environmental costs that might be incurred.[43] In the House of Commons, Eldon Griffiths (Under-Secretary at the Department of the Environment) affirmed that Britain would not be 'shoehorned' into acceptance of the Commission's proposals. Joining the EEC did not mean that we would be pressed into increases. He said that the Transport Minister would press very strongly on the Community 'the merits of restraining the increasing size of lorries in the interests of the environment and the comfort of all people in Europe'.[44]

There were, however, pressures disposing the Six to reach the earliest possible agreement. Given that the other EEC candidate members were broadly in agreement with Britain on the gross weights issue and had axle limits of 10 tons or less, a sizeable opposition bloc could be formed against the Commission's proposals, if a decision were to be delayed until after Community enlargement. With this in mind some nations, particularly Italy and France, were eager to achieve compromise quickly in an effort to minimize the concessions which they would have to make. The weights issue was also seen as a test of the 'political will' of the Six on a question which has great priority in terms of Community principle, but on which little progress had effectively been made. Community members on both sides of the weights issue felt, therefore, that a decision should be reached without the benefit of outside interference.

In an attempt to secure a unanimous agreement, at the next meeting of the Transport Ministers of the Six in May 1972, Marcel Mart (the Luxembourg Minister presiding over the meeting) put forward a compromise axle weight figure of 11 tons together with an eight-year adoption period to accommodate the interests of the French lorry manufacturers who are geared to the production of 13-ton axle weights. Prior to this meeting, the British Transport Minister, at a meeting with Albert Coppé (the EEC Commissioner responsible for Transport) had argued forcibly for the maintenance of the 10-ton axle limits.[45] On the eve of the ministerial meeting, Peyton again spoke out strongly against further weight increases, this time at a luncheon in Brussels. It would, he argued, 'be an unwarrantable intrusion on the quality of life if over-large lorries were allowed to force their way through city streets and country roads'.

Despite these efforts the EEC Ministers resolved the dispute and 'agreed in principle' to an upper gross operating weight of 40 tons and the lower compromise figure of 11 tons for axle loadings. Both sides were satisfied with provisions stating that these would not come into effect until 1980 and that existing national limits on weights and dimensions would be allowed to continue for domestic traffic until 1985.

Because of the Community's impending enlargement and the objections tabled by the prospective members, the Council of Ministers decided to submit its compromise to the candidate states for their consideration, and to meet again (possibly augmented by Ministers from the new member countries) to ratify their agreement. The Ministers of the Six had obviously agreed that it would appear 'sharp practice' to present the candidate countries with a *fait accompli* so near to their full membership, and thus, to avoid creating an unpleasant atmosphere, had decided not to adhere to a strict procedural line. Immediately following this decision, John Peyton formally requested consultations with the Six at ministerial level (rather than at the permanent official and ambassadorial levels at which consultations are usually conducted) to re-emphasize Britain's case for lower limits.

To the great satisfaction of environmentalists, the Minister vigorously pursued his efforts to prevent a binding decision being taken by the Council of Ministers. Ignoring a report by the EEC Commission which had recommended that representations by candidate member countries should be rejected, John Peyton reaffirmed (at a meeting with European Transport Ministers in November 1972) that the economic and environmental costs of increased lorry weights were too great for Britain to incur. He also reminded his audience that the political cost (in terms of sympathy for European integration) of imposing unwanted regulations on candidate members, might bear heavily on the EEC itself.

Although privately British representatives to the EEC had agreed in principle to the higher gross operating figure (while insisting that there could be no concessions on axle weight) the British Transport Minister refused any premature consideration of a compromise formula,[46] and secured from the Council of Ministers an assurance that no binding directive would be used until such an agreement had been reached.

In the period before the next ministerial meeting, developments in Parliament did much to reinforce the determination of the British Government. An Early Day Motion tabled by Angus Maude (Cons., Stratford) urged the Commons 'to reject any proposals to increase the maximum axle weight and size of lorries', and received the signatures of over 100 MPs from all parties. In an attempt to capitalize on this situation and inflict on the Government another defeat on the Common Market issue, the Opposition tabled a similar motion for debate on one of their allotted days. In the debate any increase in lorry weights was almost unanimously opposed and several Members even indicated tacit support for direct action by those communities with particularly severe lorry problems. The Government, although complimented for its defensive stand in Europe, was criticized for giving too much emphasis to the question of axle weights, while playing down the equally vital question of gross weights. Since political principle and government self-interest coincided in this situation, there was little surprise when the Government gave its qualified support to the Opposition motion, which was then carried without a division.[47]

European Ministers were left in no doubt regarding the intensity of British feeling on the 'weight' issue, when John Peyton again presented the case for the postponement of a decision at a meeting of European Ministers in December. The Transport Ministers, having been subjected to considerable political pressure finally capitulated and agreed to Britain's proposal.

For those who had been campaigning for almost four years this interim solution was received with great relief, and encouraged hopes for a favourable settlement in the future. For a time, it had seemed that the victory achieved by the environmental movement in 1970 had been in danger of reversal with the location of responsibility seeming to be out of reach of protesters in Britain. In the face of this threat, several of the amenity bodies extended their campaign into Europe. The CPRE, in an open letter to Albert Coppé, voiced 'on behalf of the majority of the inhabitants of Great Britain' their total opposition to the Commission's proposals, and called for a fundamental rethinking of transport policy to give proper consideration to environmental needs.[48] It has also called on its branches and constituent and affiliated organizations to enlist support from European amenity societies with which they are in contact.[49] Similarly the RIBA has appealed to European professional organizations for their help in what has now become an international campaign.

For the most part, however, the efforts of amenity societies and of local government have been directed at the British Government in an attempt to encourage the Minister in his European negotiations,[50] and to press for the establishment of a system of control for vehicles presently operating at home. There was little reaction to the new regulations issued by the Government allowing intermediate weight increases for some vehicles below the 32-ton limit, and the prime concern has been to urge the introduction of designated routeing systems to exclude heavy lorries from city centres, historic sites, and other damage-prone areas. In this connection the CPRE has urged local authorities to make more use of their powers to keep heavy vehicles off unsuitable roads.[51]

The Department of the Environment has for some time been considering measures to restrict the routes available to heavy vehicles. Several times during the last two years, the Minister for Transport Industries reaffirmed his own opinion that it was essential to move towards such a policy. He is on record, however, as stating that such restrictions could not be left to local regulation but must be part of a national scheme.[52] When a department initiative on this question eventually came, in the summer of 1972, the Minister appeared to have departed from this view. Having considered the logistical complexity of drawing up a national scheme, and the possible consequences for hauliers and the existing distribution networks, the Minister had opted for a somewhat attenuated restriction policy. The draft circular issued to local authorities and the amenity societies merely proposed a system of advisory lorry routes, drawn up not by the central Government but by individual local authorities. These proposals described by the Labour Party spokesman on transport as 'a sop to the environmentalists' were received with little enthusiasm by the local government organizations and amenity societies alike. While welcoming them as a first step in the right direction their verdict was that such advisory schemes were unlikely to have much influence and were no substitute for a mandatory system. Representations to this effect, however, seemed to have little influence upon the final content of the circular.

Several local authorities have already drawn up proposals for restriction schemes. For example, Derbyshire County Council (in consultation with the Peak Park Planning Board) has issued a blueprint for a 3-ton weight limit in a forty-square-mile area of rural Derbyshire.[53] Similar proposals have been made by two north-Westmorland highway authorities for twenty miles of A-class road, which in the past have suffered severe damage. Colchester Borough Council has actually introduced an experimental scheme banning heavy vehicles from the town centre and imposing a £50 fixed penalty for offenders. The Greater London Council has announced that lorries greater than 40 ft in length are to be banned from central London. This decision was taken without consultation with the road-haulage industry, however, and following strong representations from the RHA, the scheme has been modified to allow access to those vehicles with depots in the affected areas. It is significant that the Secretary of State for the Environment has circularized all local authorities, requesting that they consult with the RHA before drawing up restriction plans.[54] The importance of the above schemes lies in their mandatory character and the legal penalties to which failure to comply would bring. They constitute an effective step (if taken) towards the implementation of the Marshall Committee's recommendations, but concomitantly they would impose on hauliers longer journeys and increased costs, which would undoubtedly result in higher prices for consumer goods. On the other hand the advisory system proposed by the Government leaves much to the discretion of operators and drivers, and therefore requires a great deal of good will and social conscience to be effective.

It appears, however, that hauliers may not be required to shoulder such responsibilities, for only one month after the publication of its circular on advisory routes the Government was persuaded to consider much stronger measures. A Private Member's Bill tabled by Hugh Dykes (Cons., Harrow East) originally provided for a National Advisory Council which would make recommendations for the control of lorry movements. The Bill envisaged a zonal system whereby enforceable prohibitions, restrictions, or advisory controls would apply to heavy vehicles and which would come into effect by 1976. Qualified support came from the Government during the Second Reading, though it was suggested that the new county authorities might be better placed (than an Advisory Council) to formulate such proposals, whilst still working within the timetable envisaged by the Bill.[55]

Realizing that strong local authority opposition to any suggestion for a National Advisory Council is the limiting factor determining the extent to which the Government is prepared to go on this issue, the amenity societies warmly welcomed these proposals as a recognition of the principle for which they have been campaigning over the last few years. This does not mark the end of their campaign, however, for several of the organizations which initiated the controversy over the 'juggernauts' now see this issue as merely one part of a general transport problem and of an impending transport crisis which cannot be solved by piecemeal remedial action. The campaign has now broadened to press for the formulation of a comprehensive, fully integrated and environmentally sound transport policy, with the object of reducing not only the environmental damage caused by road vehicles, but also their significance (in relation to other forms of transport) in the economy as a whole.

## Notes

1 The authors would like to thank Mr Arthur Percival of the Civic Trust, Mr W. Mills of the Freight Transport Association, Mr J. D. W. Gent, Deputy Director of the Society of Motor Manufacturers and Traders and Mr W. W. Shiers of the DOE, for their co-operation and assistance in the preparation of this article.
2 Civic Trust report, 'Heavy lorries', October 1970, p. 32.
3 See *Motor Transport*, 18 July 1969.
4 Ibid.
5 See memorandum, *The Case for Increased Vehicle Weights*, 10 December 1970, from the SMMT to the Minister for Transport Industries. This summary of the main arguments was presented to the Minister only a few days before a decision was taken on the weights issue.
6 Ibid., p. 2.
7 Ibid.
8 'Case for heavier vehicles', see *Guardian*, 18 September 1970, for a summary of the haulage industry's case.
9 See *Motor Transport*, 18 July (also confirmed by spokesman of the Department of the Environment in interview, March 1972).
10 See *Sunday Times*, 3 August 1972, Tony Dawe, 'Fear of the 40-ton lorry'.
11 See *Annual Report* of Faversham Society, 1968-9, pp. 2-3.
12 The Joint Committee, composed of representatives of the four societies, was formed in the belief that the societies would speak with greater impact if doing so jointly. All important cases and matters of principle affecting legislation are referred to it by the constituent bodies. Its meetings are attended by representatives of other interested amenity bodies. (See *Report of Victorian Society*, 1969-70.)
13 See *Annual Report* of Georgian Group, 1969, pp. 8-9.
14 The signatories included: the Bishop of Chester, President of the Pedestrians' Association for Road Safety; Lord Euston, Chairman of the Society for the Protection of Ancient Buildings; Lord Kinnoull, Chairman of the National Council on Inland Transport; Graham Page, Esq., MP, Treasurer of the Pedestrians' Association for Road Safety.
15 At a later stage in the campaign, during the 1970 election period, the Pedestrians' Association sent out an appeal to all parliamentary candidates urging them to oppose heavier goods vehicles.
16 See report of public meeting, *The Lorry and the Community*, published by Pedestrians' Association for Road Safety, 13 November 1969.
17 Summary of discussion in Parliament reported in *Roadway*, January 1970, 'Both Houses oppose heavy lorries'.
18 See *Sunday Times*, 4 January 1972, 'Big lorries cost 10 million per year in road delays'.
19 Notes of meeting on Proposed Increases in Weights of Goods Vehicles held on 10 February 1970 at Ministry of Housing and Local Government.
20 Notes of meeting held on 6 March 1970 at Scottish Development Office.
21 CoEnCo press release, 30 January 1970. The sub-committee was chaired by Peter Robshaw, a member of the Civic Trust.
22 In fact in June the Minister commissioned a report along these lines from the Road Research Laboratory, but it was not completed before the Minister arrived at his decision in December 1970. See *Roadway*, June 1970.
23 See William Plowden, 'The roads lobby' in A. Barker and M. Rush, *The Member of Parliament and his Information*, p. 89, Allen & Unwin, 1970.

24 Road Research Laboratory *Report* L316, 1970, 'A study of fatal injuries in vehicle collisions, based on Coroner's Reports', reported in *Sunday Times*, 24 May 1970.
25 Road Research Laboratory *Report* LR357, 1970, reported in *Roadway*, December 1970.
26 Road Research Laboratory Annual Report for 1969 reported in *Roadway*, December 1970.
27 See two articles by Graham Jenkins, BA, FCIS, Chairman of the Pedestrians' Association for Road Safety: (1) 'Heavier lorries?' in *Housing and Planning Review*, July-August 1970; (2) 'Transport and the environment' in *Housing and Planning Review*, July-September 1971. These articles present a broad summary of the case against heavy lorries on both environmental and economic grounds, based on the evidence from these and a number of other governmental reports.
28 White Paper, *Reorganization of Central Government*, Cmnd 4506.
29 The authors are currently engaged in an analysis of parliamentary Questions concerning 'the environment'. Preliminary results suggest that a large proportion of these Questions are directed at departments other then the DOE.
30 See *Roadway*, journal of Road Haulage Association, August 1970.
31 See CoEnCo Annual Report, September 1970.
32 Barbara Maude (wife of Angus Maude, MP), now Chairman of the Transport Reform Group and Chairman of the Oxfordshire branch of CPRE, stated in September 1970 that 'so many people were affronted by the intrusion that heavy traffic made on their lives that if the position worsened they would stage sit-ins on the road'. *Sunday Times*, 20 September 1970.
33 Letter to *The Times*, 25 September 1970.
34 CPRE press release, 26 October 1970.
35 See Tony Aldous, *Battle for the Environment*, Fontana, 1972, p. 65, 'Lorries and towns'.
36 Department of the Environment press release, 16 December 1970.
37 In April 1972, Peter Walker again exerted his authority over one of the Ministers. Following a statement by Graham Page (Minister for Housing and Local Government) regarding proposals to release more land for housing development, the Secretary for the Environment publicly denied that such measures would be undertaken.
38 See *Annual Report and Accounts* of Freight Transport Association, 1971-2, p. 9.
39 Paper given by M. J. W. Furness at Road Haulage Association Conference, 1971.
40 See statement by E. T. Twemlow, Managing Director of Fodens Ltd, in *The Times Business News*, 1 November 1971, 'Truck weight verdict key to sales'. But also see *Financial Times*, 7 June 1971, 'Lorry market slows to crawl' for a deeper analysis of problems faced by the truck industry.
41 See *New Society*, 13 April 1972: 'The 44 tonners' by David Starkie, pp. 63-4. David Starkie is Lecturer in Transport at the University of Reading.
42 See *European Community*, January 1972, 'Eurolorries'. (References to weights proposed by the EEC Commissioner, and discussed by the Council of Ministers are in fact in metric tonnes, 1 imperial (British) ton = 1·016 metric tonnes.)
43 *Sunday Times*, 30 April 1972, editorial, 'Eurojugger'.
44 See statement by Eldon Griffiths quoted in the *Guardian*, 3 February 1972, 'Curbs on heavy lorries will stay'.

45 It would appear that John Peyton argued 'too forcibly at this meeting and the parties did not bid farewell amicably'. See *The Times*, 22 May 1972, 'Business diary in Europe'.

46 *Daily Telegraph*, 7 November 1972.

47 *Hansard*, 29 November 1972, cols 511-60. 'Early day motions' are those tabled by backbench MPs which appear on the Order Paper for discussion on 'an Early Day'. An Early Day seldom arrives, however, and the motions remain undebated expressions of backbench opinion. See S. E. Finer *et al.*, *Backbench Opinion in the House of Commons*, Pergamon Press, 1961.

48 Letter from A. Holford-Walker (Secretary, CPRE) to Albert Coppé (European Commissioner for Transport), 9 October 1972. Reprinted in CPRE *Bulletin*, August-September 1972.

49 Letter from A. Holford-Walker, 10 October 1972.

50 Letter from four local authority associations (County Councils Association, Rural District Councils Association, Urban District Councils Association and Association of Municipal Corporations) to John Peyton, 28 November 1972. The associations have also voiced their views on the weights issue through the European Joint Group – the Government/local authority forum established for the discussion of matters arising from British entry into the EEC.

51 CPRE Annual Report, 1971.

52 See *The Times*, 18 January and 15 June 1972.

53 See *Routes for People, an Environmental Approach to Highway Planning*, report by Derbyshire County Council and the Peak Park Planning Board, 1972.

54 See *Roadway*, journal of the RHA, March 1973.

55 *Hansard*, 2 February 1973, cols 1786-850.

# Chapter 8

# The Roskillers: Cublington fights the airport[1]

Richard Kimber and J. J. Richardson

O Heavenly Father, we beseech Thee, to behold in Thy power and mercy Thy people in this village, and in the other villages roundabout us, who are at this time in great distress at the threat that their houses of God, the resting places of their dead, their homes and villages may all be destroyed. We pray Thee that the hearts of those who contemplate this, and have the power to carry it out, may be turned and that deliverance from this desolation may be granted unto us, to the praise of Thy Holy Name.

Extract from a prayer composed in 1970 by the Rector of Hoggeston with Dunton

**The background to the Roskill inquiry**

The Government's announcement on 26 April 1971 that London's third airport was to be sited on the Essex coast at Foulness marked the end of a protracted debate which began in 1953. It was in July of that year that the Conservative Government decided that Gatwick should be developed as London's second major airport, but that Stansted in Essex should meanwhile be retained as a reserve airport until Gatwick was sufficiently developed. At that time Stansted was considered to be unsuitable largely because it was too far from London. However, by 1961 it had become clear that London might have need of a *third* airport due to the unexpectedly large increase in air traffic and, prompted by a Report from the House of Commons Select Committee on Estimates,[2] it was decided to institute an interdepartmental committee of inquiry into the whole problem. The report of the committee was published in

1964 and it recommended that Stansted should be the site for the third London airport. This recommendation was accepted by the Minister of Housing and Local Government, Richard Crossman, and an Inspector was appointed to hold a public inquiry into 'local objections relating to the suitability of Stansted for an airport, and the effect of the proposed development on local interests'.[3] The fate of this public inquiry (which lasted from December 1965 to February 1966) caused considerable bitterness in the Stansted area, for not only was publication of the Inspector's Report delayed until 1967, but his findings were eventually rejected by the Labour Government.

The findings were in fact very clear, for the Inspector considered that 'it would be a major calamity for the neighbourhood if a major airport were placed at Stansted. Such a decision would be only justified by national necessity. Necessity was not proved by evidence at this Inquiry.' On the face of it, this should have been splendid news for the people living in the Stansted area but on receiving the Report the Government decided to institute its own internal and secret review of the whole problem. The conclusions (though not the evidence) of this internal review were published in May 1967 in the form of a White Paper.[4] The Government decided that, despite the Inspector's Report (published simultaneously with the White Paper), Stansted was the most suitable of all likely sites for the airport. The decision to publish both the Inspector's Report and the White Paper opposing it on the same day is difficult to explain. Cashinella and Thompson[5] suggest that it was an attempt to present the country with a *fait accompli* and that Mr Douglas Jay (President of the Board of Trade), in selecting the date of publication, took special note of the fact that the House of Commons was about to begin a three-week recess. Whether or not this was deliberate, it certainly failed to stifle Parliamentary criticism since Mr Jay received a singularly hostile reception when the White Paper was debated in the House. The Conservative Opposition put down a Motion calling upon the Government to set up an independent committee of inquiry into *national* airport policy, in the context of which a decision on a third London airport could eventually be taken.[6] In moving the Motion, Robert Carr argued for an independent, highly professional and comprehensive inquiry, and expressed the almost prophetic view that 'no Government should high-handedly override a public protest which is as intense and widespread as it is in this case'. Support for the Motion was not confined to the Conservative side of the House; both Mrs Renée Short (Wolverhampton North-East) and Mr Stan Newnes (Epping) criticized the decision and refused to support their Government in the division. In particular, Mr Newnes argued that it was clearly part of a 'very effective public relations job' to suggest that all aspects had been taken into account. This was a reference to the fact that the British Airports Authority, which favoured Stansted, used a subsidiary of the London Press Exchange as public relations consultants. It is interesting to note that two former Labour Party employees were handling the BAA account at that time.[7] As we shall see later, this was not the last occasion on which the use of public relations techniques in the airport saga was criticized.

Despite cross-bench criticism in the debate, the Government had a majority of sixty-four in the Division on the Opposition's Motion and its own amendment, reaffirming the decision to select Stansted, was carried by a similar majority. That the Government had to resort to the use of a three line Whip to maintain its majority is evidence of the strength of feeling on the Labour benches against the Stansted decision. An even clearer indication of Commons' opposition had been given earlier in June when over 230 Members signed an Early Day Motion requesting that a Royal Commission should be appointed to investigate the whole problem before an expansion of Stansted Airport was agreed. With such firm Parliamentary opposition to the selection of Stansted without any really wide-ranging public inquiry, the Government could ill-afford to face a vociferous campaign against the decision outside Parliament. Yet this is just what did emerge, in the form of the North-East Essex and East Hertfordshire Preservation Association led by Mr John Lukics (a local councillor and farmer) and Sir Roger Hawkey (a businessman). The Preservation Association had been active at the original local public inquiry at Stansted and eventually managed to raise £25,000 from over 13,000 members.[8] When they heard that the Government had, despite the Inspector's Report, chosen Stansted, they intensified their campaign.[9] It is fair to say that this, combined with the Parliamentary pressure, finally convinced the Government that some form of investigation which would be seen to be fairer than the secret internal review was necessary. Consequently, on 22 February 1968 the new President of the Board of Trade, Anthony Crosland, announced that the whole matter would be re-examined. The Minister explained that the decision to hold a new inquiry was made necessary because discussions had been held with local authorities and other bodies about the realignment of the proposed runways at Stansted, with a view to reducing the number of people likely to be affected by noise. Since these consultations had resulted in new alignments which would affect people who had not objected at the original inquiry, and since the new proposals were for a *four*-runway airport (instead of three), they 'amounted to such a radical departure from earlier proposals as to constitute a virtually new project'. In announcing the new inquiry, Mr Crosland freely admitted it was worth taking a serious risk of delay 'in order to satisfy public opinion that the right decision is taken and that the matter has been examined objectively'. He had reached this conclusion in response not merely to the very legitimate pressure of the House, 'but to the clear signs both locally and nationally, which have emerged over the last few months that public anxiety and debate about the decision continues'.[10] It was no doubt much easier for Mr Crosland, as the new Minister, to reverse Mr Jay's earlier intransigence, but in doing so the Government had in effect given notice that a campaign of the Stansted type could significantly influence government thinking. As one commentator argued three years later, the uproar over Stansted indicated that any inland site within reach of London might prove a *political* impossibility.[11] Indeed, Mr Crosland, though he subsequently criticized the Conservative choice of Foulness,[12] did agree in February 1968 that the possibility of an offshore site must be within the terms of reference of the new inquiry.[13]

## The Roskill Commission

Three months later, on 20 May 1968, Anthony Crosland announced to the House that a non-statutory Commission was to be set up under Mr Justice Roskill with the following terms of reference: 'To inquire into the timing of the need for a four-runway airport to cater for the growth of traffic at existing airports serving the London area, to consider the various alternative sites and to recommend which site should be selected.'[14] The Commission were given certain specific directives by the Minister. They were asked to look at general planning issues, including population and employment growth; noise; amenity and the effect on agriculture and existing property; aviation issues, including air traffic control and safety, surface access; defence issues (i.e. they had to look specifically at the problem of the Shoeburyness firing range when considering Foulness as a possible site); and cost, including the need for cost-benefit analysis. In addition, the Government laid down the broad procedure to be followed by the Commission in conducting the inquiry. It is worth dwelling a little on the procedure at this stage as it provided the formal framework within which the campaign against siting the airport at Wing/Cublington had to be fought. Basically, the inquiry had to proceed in five stages. Stage I was to be the general sifting of all possible sites in order to produce a short-list for more detailed consideration. At this stage there was to be no right of representation before the Commission. At the end of Stage I the Commission would announce its short-list of sites including a definition of their approximate boundaries including flight paths and runway alignments. This was intended to enable those living in the area to understand how they would be affected. Stage II was to consist of the local hearings at each short-listed site and to be conducted by a senior planning inspector (who would also be a member of the Commission). As with all local inquiries, interested parties were entitled to be represented by legal and other experts at this stage. It was recognized at the outset that this might prove a lengthy process and so Stages III and IV were to progress alongside Stage II. Stage III was to be an investigation into 'matters relevant to the choice to be made between the sites' (including surface transport, noise, regional planning, etc.). This work would be conducted by the Commission's staff and consultants, and would include technical evidence submitted by interested parties. Stage IV was to be a consideration by the Commission of all the material produced in Stage III. Stage V was to be the final public hearing where interested parties, again with legal/expert representation if they wished, could test and challenge material produced in earlier stages. At the end of this lengthy process, expected to take two years, the Commission would prepare a report for the Government. The names of the seven Commissioners were announced on 24 June 1968. Chaired by Mr Justice Roskill, they were Colin Buchanan, Professor of Transport at Imperial College of Science and Technology and author of the well-known *Traffic in Towns*; A. Goldstein, a partner in R. Travers Morgan and Partners, consulting engineers; A. J. Hunt, Principal Planning Inspector, Ministry of Housing and Local Government (later made part of the Department of the Environment); David Keith-Lucas, Professor of Aircraft Design, Cranfield Institute of Technology; Arthur Knight, Deputy Chairman of Courtaulds Ltd; A. A. Walters, Cassel Professor of Economics, University of London.

## The origins of the Wing Airport Resistance Association

During Stage I of the Commission's inquiry a list was compiled of some seventy-eight possible sites, of which Silverstone in Northamptonshire was considered to be one of the more likely locations – even though the area was shown on the County Development Plans as an area of great landscape value.[15] It was not surprising, therefore, as word spread that Silverstone might be short-listed, that concern should develop in that area. Desmond Fennell, a young barrister who had recently settled at Winslow (about ten miles from Silverstone), embarked on a campaign to prevent the selection of Silverstone as the airport site. In the Autumn of 1968 Fennell and his wife together with Nevile Wallace, whom Fennell had met shortly before at a planning inquiry, began to contact local organizations with a view to convening a meeting on 9 November to discuss what action should be taken. Also involved in the birth of the resistance movement was Derek Lawson, a solicitor from Towcester who was able to help with office space and to provide useful contacts through his position on the local Executive of the County Landowners' Association (CLA). It was the good fortune for what was eventually to become the Wing Airport Resistance Association (WARA) that these three individuals, all with legal training and at least two of them with considerable experience of planning matters, should have joined forces to form an *ad hoc* committee to resist the proposal of Silverstone as the airport site. Nevile Wallace's contribution was particularly important in these early stages; as an official of the National Farmers' Union he was an expert on land use and planning inquiries of all kinds, and had been involved with the Stansted inquiry.

However, while this group of 'young professionals' commanded considerable expertise and could organize opposition on a technical level relatively easily, they faced the much more difficult *political* problem of demonstrating that genuine 'grass roots' opposition existed within the local communities. Indeed some of the earliest signs indicated that widespread opposition to an airport at Silverstone might be slow to materialize. For example, in response to a letter from Fennell inquiring about local feeling, a Northampton solicitor reported that the Northamptonshire County Council would do little to object until and unless Silverstone was short-listed. It even appeared that some local residents might welcome an airport since it would put Silverstone on the map, and they appeared not to be too bothered about aircraft noise as they were quite used to the noise from the Silverstone motor racing circuit![16] The most encouraging response, from Fennell's viewpoint, was that the agricultural community was bitterly opposed to the airport and a meeting of the local NFU had 'strongly objected' when told that upwards of 7,000 acres of land might be required.

At the meeting on 9 November, held at Lawn House, Winslow (Fennell's home), it was decided to lobby members of Buckinghamshire County Council. It appeared that this Council was more inclined to fight the airport and it was hoped that the County Councillors could be persuaded to support their own planning committee in recommending that the County should oppose the selection of Silverstone. A second meeting was held at Lawn House on 30 November 1968 to which a cross-section of organizations in the area were invited. Members of the County Planning Offices agreed to attend, as did representatives of the County and local branches of the NFU, local branches of the CPRE and the CLA (represented by Lawson), National Union of Agricultural Workers, Stowe School, Silverstone Circuits, and many Parish, Rural District and County Councillors.[17] In all, approximately fifty people attended and it was decided to form an Executive Committee composed of representatives of the organizations attending and members of the general public. Fennell was elected Chairman, Wallace as Deputy Chairman, and Lawson as Secretary/Treasurer, with twenty or so other people representing the various interests present. The Committee was charged with considering how the campaign might best be organized and was asked to report back in due course. This it did on 8 February 1969.

Meanwhile the County Councils were organizing their own opposition. A meeting was called jointly by the Buckinghamshire, Northamptonshire, and Oxfordshire County Councils at Brackley Town Hall on 19 December to which representatives of Urban and Rural District Councils, the NFU (represented by Wallace), the National Union of Agricultural Workers, and various established amenity and preservation societies in the area were invited. That the newly formed Committee was not initially asked to send a representative illustrates one of the major early difficulties they faced – that of establishing their credibility. However, after one of the invited representatives withdrew, Lawson was allowed to attend as spokesman for the new committee (no doubt after some prompting from Wallace). This was an important stage in the organization's development, for it meant that it had been recognized as a legitimate expression of opinion in the area and could be encouraged to submit evidence to any local inquiry which might be held.

By February the Executive Committee was ready to report the results of its deliberations. It had chosen a name for the organization – the Silverstone Airport Resistance Association (SARA) – and had drawn up a constitution. It had also given some thought to its two main problems: raising funds and establishing a reputation as a responsible body whose opposition to the airport should be taken seriously. At a fairly early stage Lawson, as Secretary/Treasurer, decided on a novel method of raising funds for the Association. He had been concerned that they might raise a large sum of money 'without any clear idea what use might be made of it', and he decided to approach individuals not for donations but for a *promise* of a donation if Silverstone was short-listed. By February, using this system of 'promissory notes' Lawson had over £800 in guarantees for the Association. Of course, money was required for current activity, particularly for printing the leaflet *Silverstone and You*, which attempted to set out the implications of siting a major airport in the area, and which advocated Foulness as a better site. In an attempt both to bring money into the Association and also to give it further solidity and respectability, a number of prominent people were asked to associate themselves with SARA by becoming Vice-Presidents. Some local MPs (Arthur Jones, South Northants.; Neil Marten, Banbury; Sir Spencer Summers, Aylesbury; Robert Maxwell, Buckingham) had already been sounded out by Mrs Fennell in early November, when it had been realized that support at a higher level might be required. So that the MPs should not dismiss the resistance organization as 'cranks', Mrs Fennell also wrote to an MP, for whom she had once worked, asking her to lobby them on the matter and stressing the urgency of the situation, 'no one seems to realize how desperate the position is and Stansted will win because nobody in North Buckinghamshire appreciates what a jumbo jet is!'[18] Subsequently, two of the MPs approached, Sir Spencer Summers and Arthur Jones, agreed to become Vice-Presidents and they were joined by Lady Hesketh, a landowner, Alderman D. E. Hutchinson, Chairman of Northamptonshire County Council and Major General Vivian Street, the distinguished soldier, who – in addition to being a Steward of the Jockey Club – was well known for his charitable work particularly with the Save the Children Fund. Mr David Robarts, a landowner and Chairman of the National Provincial Bank, became the titular head of the Association as its President.

While SARA was being established on a formal basis, the County Councils (through their joint committee established at Brackley in December) were encouraged to press the Ministry to postpone Stage II of the Roskill investigation in order that more time might be made available to engage experts and to prepare evidence. At the same time both the CLA and the CPRE became more active. The CLA formed a joint co-ordinating committee, under Lawson's chairmanship, to organize opposition in Buckinghamshire, Northamptonshire and Oxfordshire, and the CPRE set up a similar body on which SARA was also represented. Thus at an early stage a considerable degree of overlapping membership was established between the various official and voluntary organizations which were opposed to the airport. This pattern persisted when SARA was subsequently transformed into WARA and was of considerable value to the campaign. Nevile Wallace, in particular, was able to make an important contribution because he always acted in a dual capacity (i.e. both for SARA and the NFU) in dealing with the County Councils and the Roskill Commission. In this way he was able to use his contacts to obtain useful information. For example, he managed to establish that if Foulness was not selected, Essex County Council would strongly advocate Silverstone as the best site.[19]

As SARA developed, its reputation as a viable grass roots

organization increased and in February Lawson wrote to the Clerk of Northamptonshire County Council suggesting, in detail, the role which SARA might play. It was agreed that SARA should represent the County Council in the presentation of 'certain angles for objection', although at that stage the Council was not clear exactly which angles could be left to SARA – probably because the three County Councils were finding it difficult to evolve a common policy. Also, at this stage SARA itself had not really decided whether or not it should be a mass movement, but Lawson correctly realized that if SARA was to play a much wider role than that envisaged by the Councils, then this could only be achieved by opening up the membership of the Association to all and sundry 'in the hope that we will obtain sufficient members to enable us to claim that we represent a substantial section of public opinion'.[20] In retrospect, it is doubtful if SARA and WARA could have avoided such a development for by February 1969 a number of existing voluntary organizations were looking to SARA for assistance. These included such diverse organizations as the Racehorse Trainers and Bloodstock Breeders, the local Conservative Association, local branches of the CPRE, local Women's Institutes and a local company, the Deanshanger Oxide Company. In addition there were signs of the emergence of spontaneous grass roots activity when a branch of SARA was formed at Brackley without the knowledge of the central SARA Committee. However, before these developments could be rationalized, the organizers were overtaken by events. On 4 March 1969 the Roskill Commission published its short-list of four sites *excluding* Silverstone (and, to everyone's surprise, Stansted too). On the face of it, this should have marked the end of SARA and technically it did. In practice it merely marked the evolution from a small, localized, resistance movement to the wider mass movement known as WARA.

## Wing/Cublington:[21] the nature of the community

The Roskill Commission published its short-list of sites on 4 March 1969. They were Foulness in Essex, Thurleigh in Bedfordshire, Nuthampstead in Hertfordshire, and Cublington (Wing) in Buckinghamshire. Fortunately for historians, the Commission eventually engaged a team of sociologists (from Essex University) to design a survey of the four short-listed sites to be carried out by National Opinion Polls, as part of the Stage III inquiry. What is remarkable is 'the extreme time constraint', as the Essex team put it, under which the survey report was produced.[22] Why the decision to commission the sociological field work was made so late is not clear. In the event, the objectors at Wing, although not uncritical of the survey, had reason to be satisfied with some of the findings. Indeed, this presented some WARA members with a dilemma: while wanting to draw attention to the study's shortcomings, they also wished to support its main findings. This problem led to 'a certain amount of internal politics',[23] the net result of which was that criticism of the survey did not play a major part in the campaign.

Perhaps the most interesting finding was that, contrary to popular belief, Cublington was not more middle class than the other inland sites. The Essex team found that, leaving Foulness aside (with 83 per cent of its population being semi- or unskilled manual workers or pensioners), 'Thurleigh has the highest proportion of non-manual workers (51 per cent) and lowest proportion of manual workers (49 per cent), Cublington the lowest proportion of non-manual workers (40 per cent) and highest proportion of manual workers (59 per cent). In this sense (and this is reinforced by the tenure and income data) Thurleigh is the most "middle-class" of the three inland sites and Cublington the least; or put the other way, Cublington is the most "working-class" of the inland sites and Thurleigh the least. Nuthampstead occupies a position between the two.' Similarly at Cublington, according to the survey, only 9 per cent of those respondents who agreed to answer the question had incomes above £1,500 per annum whereas at Thurleigh and Nuthampstead the figures were 14 per cent and 15 per cent respectively. Again Cublington had the lowest percentage of respondents who had continued their formal education beyond the age of sixteen (15 per cent compared with 23 per cent for Thurleigh and 22 per cent for Nuthampstead. On the other hand Foulness, on this as with most other factors, ranked lower than the inland sites at only 2 per cent). Nevertheless, by *national* standards Cublington had a high proportion of upper-middle and middle-class residents, as can be seen in Table 8.1.[24] Of more significance to WARA's

TABLE 8.1 *Residents in Cublington (percentages)*

| | Upper-middle and middle class (AB) | Lower-middle class (C1) | Skilled working class (C2) | Lower-working class (DE) |
|---|---|---|---|---|
| Cublington | 21 | 19 | 30 | 29 |
| National | 12 | 22 | 37 | 29 |

campaign was that Cublington was found to have a more closely knit social network than the other inland sites. For example 57 per cent of Cublington respondents had relatives living within their own parish, compared with 40 per cent for Thurleigh and 42 per cent for Nuthampstead. Similarly, 62 per cent of respondents had most of their friends living within the parish compared with 53 per cent and 56 per cent for Thurleigh and Nuthampstead. Furthermore, 42 per cent thought that their friends knew each other compared with only 28 per cent and 22 per cent respectively for the other two inland sites. (But again Foulness was the glaring exception as, although it returned a similar percentage to Cublington for 'most friends living in the same parish', some 58 per cent thought that their friends knew each other.) In terms of participation in existing local voluntary associations, Cublington also rated higher than Thurleigh and Nuthampstead (and Foulness) having a higher proportion of its population as members of voluntary associations. This was also true of membership of the airport resistance associations which were eventually formed at all four sites. At the time of the survey (August 1969) 22 per cent of respondents claimed to be members of a resistance association at Cublington, compared with 17 per cent for Thurleigh and Nuthampstead and 10 per cent for Foulness. Not surprisingly this was reflected in the attitudes of the respondents to the siting of an airport in their midst – 70 per cent being very much opposed at Cublington compared with 53 per cent and 58 per cent at Thurleigh and Nuthampstead and 60 per cent at Foulness. In terms of likely total disruption, the research team concluded that there would be less social disruption at Foulness and less to disrupt at Thurleigh and Nuthampstead than at Cublington. It is clear, therefore, that the Cublington/Wing site was, of all the sites, probably the most fruitful area in which to attempt to rally protest. It is to the attempts by Desmond Fennell and his associates to do this that we now turn.

## The Wing Airport Resistance Association: organization and membership

Shortly before Roskill's list of potential sites was officially published, Nevile Wallace, as NFU representative, heard that Silverstone would not be included but that Wing (some sixteen miles south-east of Silverstone) would be named as a possible site. This information, which was reinforced by hints in the press, caused SARA to reorientate itself towards the south. There followed a series of meetings at which SARA was wound up and wheels were set in motion to organize resistance in the Wing area, culminating in a meeting on 10 March 1969 at Lawn House. Most of those present had been actively involved with SARA, which guaranteed an extremely smooth transition from north to south. 'Wing' replaced 'Silverstone' in the name of the organization (giving the acronym 'WARA') and SARA's constitution and bank balance were adopted by the new body. Although the group essentially consisted of the same people as had been running SARA, some reorganization took place, particularly with respect to the Secretary/Treasurership. Derek Lawson, from Towcester, by then felt that the focus of activity had shifted rather a long way from his area and, while remaining with WARA, handed over his functions to two newcomers, John Pargeter and Evelyn de Rothschild. John Pargeter, a local solicitor, took over the secretaryship and at the outset agreed to provide WARA with free office space at his Leighton Buzzard premises. This office provided the essential administrative centre of the campaign for over two years and, in effect, subsidized its day-to-day running. The treasurership was taken on by Evelyn de Rothschild, a member of the well-known merchant banking family. Other campaigning organizations are rarely fortunate enough to start with such an advantage and in the years ahead WARA was to benefit substantially from Rothschild's management skills and from the financial solidity and integrity which his name represented. One of the most important tasks, that of overseeing the 'grass roots' activity, fell to Bill Manning, a farmer and stockbreeder at Wing. Manning's standing in the local community and his contacts and organizing ability proved to be extremely important, particularly in the collection of some 60,000 signatures on a petition against the airport.

The meeting on 10 March was notable for a highly professional approach to the massive problems ahead. It was attended by a representative of a local advertising agency who outlined a possible publicity campaign, including posters, car stickers, leaflets and press advertisements. These were in fact already at the proof stage. More importantly, the new committee immediately realized the likely cost of a campaign to save Wing and they agreed that a sum of £50,000 would need to be raised. This huge sum did not apparently daunt the committee members and very few pressure groups can have started life with such confidence in their ability to raise huge sums of money. This confidence was neatly illustrated by their initial decision not to acknowledge donations below £25. This decision was one of the few serious errors made by the committee in two years and had to be hastily reversed when the WARA office began to receive irate letters complaining that donations had not been acknowledged. The question of adequate legal representation was also raised at the first meeting and it was agreed that Mr Niall MacDermot, Q.C., Labour MP and former Minister of State at the Ministry of Housing and Local Government, should be retained to present WARA's case at the Stage II inquiry. They were fortunate to obtain such an eminent advocate since the Bedfordshire County Council were also keenly interested in obtaining his services. As it happens, however, Desmond Fennell knew MacDermot personally, having worked with him at the trial of the 'Great Train Robbers'. It was also agreed that a leading firm of London solicitors should be instructed and that Nevile Wallace should look for suitable planning and technical consultants.

The organizing committee's initial confidence was very quickly justified, for at their second formal meeting on 23 March the Secretary was able to report that the total of money and promises to date was approximately £25,000. Part of this money had been transferred from the SARA account, but the bulk of it had been raised in the four weeks since publication of Roskill's short-list – and a large part of that by Rothschild on the telephone to some of his friends and acquaintances. By this time an aviation expert, D. M. Lumb (who had had experience of the Stansted inquiry), and a planning expert, Professor Tom Hancock (who had been responsible for the city plan of Peterborough), had been engaged and negotiations were in hand for the appointment of a noise expert. At this stage WARA's organizational structure was somewhat vague, with little or no formal distinction between the self-appointed Executive Committee which comprised 'representatives of interested bodies and certain private individuals', and the organization as a whole, the membership of which was 'open to all who are interested in the Aims of the Association'.[25] Consequently, attendance at these early policy meetings of WARA fluctuated somewhat, according to who happened to be invited. The third meeting on 20 April was attended by over two dozen members. It began with reports on the current state of the organization, which by this time had begun to grow rapidly and had £7,300 actually in the bank plus the £25,000 promised. Manning, as Chairman of the 'Grass Roots Committee', reported that he had formed twenty-eight sections covering 132 villages and eight towns. The collection of signatures for the petition was already well in hand and, as part of the attempt to stimulate local interest, WARA agreed to supply speakers for meetings being called by the Parish Councils and various associations in the area.

At this point the discussion took a different turn when Robert Maxwell, Labour MP for Buckingham and a Vice-President of WARA, joined the meeting and made some criticisms of the way in which the Association was being run. He felt that they were making too much noise too soon, that there was a lack of communication, that the Association was not properly representative of the area and was being treated as a public relations exercise by a few private individuals.[26] After some debate Maxwell agreed to meet the Association's Officers during the following week to discuss his objections. His intervention was of considerable significance for the development of WARA, and it initiated one of the most important policy debates within the group.

Indeed, Robert Maxwell's role in the early stages of the airport controversy is an interesting one. At first he remained uncommitted and was inclined to regard SARA as somewhat hysterical. At a lively meeting in Buckingham Town Hall on 7 February, he argued that ammunition should not be fired too soon, and said that he was appalled at the amount of early opposition to a proposal that was still in the early stages. He suggested that this early opposition was causing considerable surprise at the Roskill Commission and in Government circles.[27] Yet, when the short-listed sites became known, Maxwell came out publicly against the airport at a meeting in the Wing Village Hall. More significantly, at that meeting he introduced Fennell to Evelyn de Rothschild, who subsequently became WARA's Treasurer.[28] At a later meeting held at Stewkley on 12 April, he told a large audience that he was receiving complaints that WARA was 'an exclusive Roseberry-Rothschild affair', and threatened to dissociate himself from WARA unless the Association democratized itself and had members of all three parties on its working committees.[29] However, Maxwell's criticisms performed a useful function in forcing WARA's members to reflect on both their public image and the nature of their organization.

Four members of the Committee met with Maxwell in London five days after the Lawn House meeting and heard his detailed criticisms. Essentially, he made three points.[30] First, that he should become joint President; second, that the Association was not democratic; third, that local organizations of 1,000 members or more should be asked to send delegates to WARA meetings, and he put forward a nomination from the local Labour Party. These proposals were discussed at a meeting at Lawn House on 27 April, after a caucus of the Committee held shortly before, and it was decided to confirm the election of the existing Officers. However, Maxwell's criticisms clearly had some impact and were in part acted upon. The Executive Committee was formalized and made more representative (i.e. less in the 'Roseberry-Rothschild' image) and it was decided to hold a General Meeting in the near future to report on progress and to obtain a vote of confidence for the newly constituted Committee. Some thought was given to WARA's organization, though Maxwell's delegate structure was not accepted. However, local organizations were encouraged to pass resolutions in WARA's favour, and the communication problem, which was raised again at the General Meeting, was eventually tackled by distributing information sheets through the grass roots organization. In addition to this, considerable information was disseminated through the local press, largely due to the work of John Flewin – a freelance journalist hired by WARA as Press Officer on a part-time basis.

The Executive were naturally concerned to retain good contacts with local MPs. Nevile Wallace realized, very shrewdly, that WARA could well need the support of local MPs later in the campaign and he suggested, even before Stage II of the Roskill inquiry had begun, that MPs should be kept in reserve for the final battle. In fact this was almost exactly the role which MPs played in the WARA campaign – adding status and respectability to the Association by agreeing to serve as Vice-Presidents, but not playing an active part in the formation of WARA policies while the Roskill Commission was conducting its inquiry.

By 4 June, when the General Meeting was held in Leighton Buzzard, the division of responsibilities within the Executive had been discussed in some detail and tasks allocated. The team, which was presented to the meeting and approved unanimously, was as follows:

| | |
|---|---|
| Chairman: | D. Fennell |
| Secretary: | A. J. Pargeter |
| Treasurer: | E. R. A. de Rothschild |
| Committee Members: | |
| G. I. Ginn | (Headmaster of Stewkley Primary School, Chairman of Stewkley Action Committee, and Chairman of the Buckingham Division Liberal Association) in charge of general fund-raising activities |
| A. Hugh-Smith | (a stockbroker) |
| D. Lawson | |
| A. H. I. Lonie | (a Bletchley businessman) responsible for what was termed 'crusading', i.e. follow-up activity after meetings in new areas |
| A. C. Lovejoy | (a partner in an advertising agency) in charge of publicity and public relations with the help of Lonie |
| W. Manning | Chairman of the Grass Roots Committee |
| A. Miscampbell | (a solicitor and brother of the Conservative MP for North Blackpool) responsible for arranging speakers for the many meetings held to inform and arouse the public |
| Cllr W. Randall | (Leader of the Labour majority on the Leighton Buzzard Council) |
| D. Skinner | (Secretary, Aylesbury Trades Council) charged with co-ordinating information and letters to the press in reply to adverse comments |

N. Wallace — responsible for strategy and the co-ordination of work for the hearing

Similarly the meeting approved the President (D. Robarts) and list of Vice-Presidents:

Stephen Hastings, Esq., MC, MP
The Lady Hesketh
D. E. Hutchinson, Esq., MBE (Chairman Northants. CC)
Arthur Jones, Esq., MP
Robert Maxwell, Esq., MC, MP
Gwilym Roberts, Esq., MP
Major General Vivian Street, CMG, CBE, DSO, MC
Sir Spencer Summers, MP
Sir Aubrey Ward, DL, JP (Chairman Bucks. CC)

Thus WARA had acquired legitimacy and any charge that it was run by a self-appointed clique, responsible to no one, could no longer be substantiated. Further, it was also clear that the Association's affairs were being conducted on a businesslike basis. Membership figures had risen to 30,000, there was a sound bank balance, and the legal and technical experts were already at work on the case to be presented at the Stage II hearing at Aylesbury scheduled for July 1969 (by this time the original nucleus of experts had been supplemented by R. A. Waller, a noise expert). With the addition to the Executive Committee of Lady Hartwell (wife of the Chairman and Editor-in-Chief of the *Daily Telegraph*) the WARA team was virtually complete. Its lawyers, technical experts and supporting village organizations were able to press ahead with the campaign leading up to the local hearing at Aylesbury.

By the time of the second Annual General Meeting, in July 1970, the Roskill Commission had completed the greater part of its work and WARA took the opportunity provided by the AGM to let its members hear a first-hand account of the inquiry from Niall MacDermot and Philip Otton. This was an unusual move, for Queen's Counsel rarely meet directly with their clients without a solicitor, and special permission from the Bar Council had to be obtained. At the same meeting the Officers and Committee were re-elected, with the addition of B. C. Planton from the very active Tring Committee, and three names were recommended to the meeting as potential Vice-Presidents, Sir Geoffrey de Freitas, Lord Carrington, and Mrs Okenham (Chairman of Northamptonshire County Council).

Despite WARA's democratization, it continued to be run by the same relatively small group, a state of affairs which is inevitable in any large organization.[31] Moreover, in this situation it was also inevitable that one person, Desmond Fennell, should in practice act as king-pin of the organization. Anyone who has studied the detailed WARA files cannot fail to be impressed by the sheer volume of effort which Fennell personally put into co-ordinating the whole campaign. Very few important decisions were taken without prior reference to Fennell and he was probably the only person within WARA fully briefed in all aspects of the Association's work. In addition, he gave assistance to other organizations, such as the CPRE and the National Trust, in the preparation of their evidence.[32]

**The development of mass support**

We have indicated earlier the sensitivity of leaders of WARA to the charge that the Association was little short of an upper-middle-class clique. However, this was an accusation which persisted throughout the campaign; for example when they held a mass rally at the Wing Equestrian Centre, on 10 January 1971, Fennell received a letter complaining that this suggested 'a plea on behalf of the exclusive landed gentry and members of the hunting fraternity' whereas it was 'the more or less inarticulate majority who will be turned out of their homes willy-nilly and be compelled to find new jobs and lodgings as best they can'. Although WARA took every opportunity to demonstrate that such views had little or no foundation, one of the main answers to such criticism was the work carried out by Bill Manning as Chairman of the Grass Roots Committee. At the Stage II hearing at Aylesbury in July 1969, WARA was able to announce that it had a membership of 61,766, all of whom had signed valid membership forms which were available for inspection at the hearing. The signatures had been collected by the local WARA committees appointed in more than 188 villages and fifteen towns in the area, all under the general co-ordination of Manning. He had appointed area representatives who were asked to contact helpers in every village, to ascertain from the electoral roll the number living in each village and to ensure that WARA leaflets were distributed to as many households as possible. At the same time, village committees collected signatures on a house-to-house basis, again using the electoral roll. The membership forms were then collected by the area representatives and the running total kept by Manning. WARA was not always able to get people to realize the impact which an airport in their midst would have, despite the emphasis it put on the need to inform the population. Not all the villages directly affected by the airport matched the feat of Aston Abbotts which enrolled virtually 95 per cent of the population as members of WARA. Dunton, for example, sited on the edge of one of the proposed runways had a WARA membership of less than half, and a number of villages well within the 35 NNI noise contour[33] had no local WARA committee. On the other hand over 40 per cent of the population joined WARA in more than thirty-three villages.

The door-to-door membership drive was greatly assisted by the fact that WARA managed to develop contacts with a large number of existing voluntary associations in the area. Fennell, in his evidence at the Aylesbury hearing, listed forty-six local societies which supported WARA. These included societies keenly concerned with the environment like the Brill, North Buckinghamshire, Buckinghamshire, and Oxfordshire branches of the CPRE, but also included a number of Women's Institutes, a disabled persons' club, gardening societies and even the Berkshire, Buckinghamshire, and Oxfordshire, North Road Federation of Pigeon Societies! These local societies were encouraged to submit evidence to the Roskill Commission as an indication of the deep-rooted opposition to the airport and were generally useful in stimulating local support, both practical and financial. During the course of the WARA campaign there were many instances of existing local societies helping WARA. For example, after the announcement that Roskill had selected Wing as the best site, the Committee of the Friends of the Vale of Aylesbury sent a notice to all its members encouraging them to support WARA in all its protests and to write to their MPs and to the press. Even the pigeon fanciers wrote to WARA inquiring if they could do anything to support the Association's efforts. Useful financial support quite commonly accrued from these contacts and many societies made direct donations, including such organizations as the Aston Clinton Society, the Aylesbury Allotment Association, the Aylesbury Evening Townswomen's Guild, and the Bedfordshire Old People's Welfare Council. Manning's contacts with the racing community were also utilized and, for example, the National Hunt Jockeys challenged WARA to a cricket match which added £300 to the Association's fund. In effect, WARA managed to integrate itself successfully into the existing social and organizational structure of the local community and to use existing societies and associations as a platform for its views. As part of this, no opportunity was lost to provide a WARA speaker for any local society meeting, however small. An interesting example of WARA using the existing social network was the fund-raising event, addressed by John Betjeman, at the home of Lady Hartwell at Oving. Even so, some people were offended at not being invited since, lacking close contacts with the local community herself, Lady Hartwell had relied on the Whaddon Hunt and the British Red Cross lists, which were not entirely representative of the area.[34] Since the WARA campaign was an extended one, lasting over two years, it was important that local interest should not be allowed to dwindle and it is clear that the good contacts with local societies helped to prevent this, and to reinforce the work of WARA's own local committees.

It was this work which represented the main grass roots opposition to the airport. The village committees were set up to promote local opposition to the airport and provide support for various WARA activities, but they were autonomous and were expected in addition to organize their own activities. Not all the village committees arose at WARA's instigation: both Wings Off Wing (WOW), covering the villages of Hardwick, Oving, Pitchcott, Weedon and Whitchurch, and the Stewkley Action Committee were formed independently and, though they joined WARA, they tended to pursue a somewhat independent line. The importance of these two committees was recognized by having their Chairmen (Lovejoy and Ginn respectively) on the WARA Executive. This system of autonomous committees, while convenient from some points of view, was potentially dangerous, in the sense that acts which were harmful to the campaign might have been committed in WARA's name but without WARA's knowledge. Indeed, the Stewkley Action Committee, which even presented separate evidence at the Aylesbury hearing, tended to be rather more militant than other committees and occasionally caused the Association's Executive some embarrassment. For example, their plan to demonstrate at the Aylesbury hearing caused considerable disquiet, especially since Niall MacDermot had warned that such a demonstration might be counter productive, and Pargeter wrote formally to the Chairman of the Stewkley Action Committee asking for further details so that the matter could be brought before the full Executive. In the event, the main demonstration took place on the day before the opening of the Aylesbury hearing, and received favourable publicity.

A considerable amount of propaganda was distributed by the local committees in the form of posters, stickers and leaflets. The leaflets argued concisely against the airport, stressing that it could bring up to one million people into the area by 1990, and that 20,000 people would probably find the noise of aircraft intolerable 'and have to move away'.[35] Twelve hundred people would be forced out of their homes as soon as construction started and the cost of building schools, housing, roads, water and sewage facilities would fall largely on the local population and would be 'a rapidly increasing burden on every householder'. The alternative to all this, WARA argued, was to site the airport at Foulness where the noise would almost wholly be confined to the sea. Some of WARA's leaflets were designed to answer specific criticisms of its case, such as the claim that the airport would provide good job opportunities in the area. The Association's reply to this was that there were already three jobs waiting for every man or woman unemployed in the area and that if the airport was built it would become a monopoly employer with jobs mostly demanding specialist skills. A particularly effective leaflet included a map of the area on which noise contours were superimposed; this helped to drive home the message in areas which, although directly affected, did not seem to be responding to WARA's efforts. Some areas, however, responded very slowly. As late as January 1971 the Secretary of the Chalfont Common Women's Institute wrote to Manning saying that some people in the south of the country 'do not appear to realize that the "disturbance areas" will be so close that [the airport] will affect them'. However, this was not universal and when the Roskill noise contours were published there was a marked response and membership figures rose. The distribution of leaflets was supplemented by press advertising, and a number of large advertisements were placed in the local press, and on one occasion in the London *Evening Standard*. Also, as anyone who visited it will know, the whole area was scattered with posters and anti-airport slogans. All this was further reinforced by a series of public meetings in villages throughout the area at which WARA speakers explained in detail the likely effects of having the new London airport. By the time the Aylesbury hearing had been held, in July 1969, some eighty-one public meetings had been organized.

Although at times apathy presented a problem to the WARA Executive, at the other end of the spectrum extreme militancy at times threatened to damage WARA's responsible image. Some Stewkley posters rather amusingly depicted a local resident wielding a pitch fork in angry defiance of Roskill and the Government, but in fact there were people in the area who felt that direct action and violence might be necessary if all else failed. On a number of occasions the possibility of violence was mooted and Fennell had to repudiate such suggestions very firmly in order to prevent WARA's authority from being eroded. Thus, in reply to a letter suggesting that WARA should be more militant, Fennell replied that the Association would have 'no part in direct action or terrorist methods'. He had expressed the same view in a television appearance in response to the rumour that a quantity of dynamite had been stored in the Wing area ready for the day when construction work was to start. Shortly before Christmas it was reported in the *Observer* that pamphlets had been distributed describing the manufacture and use of petrol bombs and other weapons and the construction of barricades.

These were apparently supported by, but did not emanate from, a fifty-strong 'home defence group'.[36] It is impossible to assess the truth of all the stories about what became known as the 'Dad's

Army' problem, and no doubt the legend of the underground resistance will be perpetuated and, as time passes, expanded. Nevertheless, although no actual violence took place (discounting a hoax bomb which was placed in front of the Minister of Trade's car on his visit to the airport site in March 1971) leaflets advocating violence were distributed at the mass rally at the Equestrian Centre on 10 January 1971. These were later discovered to be identical to those distributed at the Grosvenor Square riots in October 1968 and probably had no connexion with anyone in WARA or even with the local community. Some over-enthusiastic supporters burnt an effigy of Mr Justice Roskill on Guy Fawkes' night in 1969, but Alan Lovejoy immediately wrote to Roskill apologizing for this 'totally unofficial and irresponsible act'. There was some concern that WARA might not be able to control such activities, particularly at the Aylesbury hearing. WARA's London solicitors (Lovell, White & King) specifically advised against even the demonstration organized by the Stewkley Action Committee and warned WARA not to be associated with it, since 'it could get out of hand and have the reverse effect of what is intended'.[37] That a tiny minority of WARA supporters should contemplate violence was just one more indication of the strength of feeling of most of the resident population against the airport proposal. Indeed, as we have seen, the Essex survey showed Cublington to have a significantly higher proportion of people (70 per cent) opposed to the airport than the other three sites. However, few serious problems of controlling a large and diffuse organization became apparent during the two-year campaign, and WARA was remarkably successful in channelling these strong feelings into constructive opposition, including of course well-organized and well-publicized demonstrations of a non-violent kind.

The very small group advocating violence was not the only fringe element with which the WARA Executive had to contend, for there were people living in the area who *wanted* the airport at Wing. Potentially this could have been a serious threat to WARA but in the event the pro-airport movement, in the form of the Cublington Airport Supporters' Committee, failed to make any significant impact. This was patently obvious at the Aylesbury hearing where, in contrast to WARA's 60,000 members, the Supporters' Committee could only claim 600 members, and even then could not produce satisfactory documentary evidence that these people existed. Furthermore only eight letters supporting the airport had been submitted to the Commission in contrast to the great volume of anti-airport representations which WARA had organized. Had the airport supporters managed to demonstrate a really significant degree of support for the airport then matters could have taken a rather different turn when the Government came to reach a final decision. But at no time did the pro-airport organization develop into a significant force. That this was so is in part a tribute to WARA's own efforts to convince the population. Also there can be little doubt that once the WARA campaign had got under way (and we have seen that it did so at an early stage) a considerable degree of individualism, and even courage, would have been required for people to declare their support for the airport in such a closely knit society. Considerable hostility was shown to members of the Cublington Airport Supporters' Committee; some received anonymous letters, and a brick was thrown through the window of one member's house – although WARA was responsible for none of these incidents.

Despite these various difficulties, which it must be emphasized were minor ones in terms of the whole campaign, the mass-membership drive proved to be extremely successful. This was seen by WARA itself as being of considerable significance. It was almost certainly a *necessary* achievement in that WARA had to show that it had mass support since the decision was, ultimately, a political one. Even so the Executive had the good sense to realize that to demonstrate the 'popular will' might not be sufficient. Roskill had been given what was essentially a technical brief and the WARA Executive had, therefore, to fight a technical as well as a popular battle. From the outset they realized the necessity of this and from the outset they appreciated the likely cost. Consequently a considerable proportion of the Executive's efforts was devoted to the problem of raising sufficient money to finance the technical and legal representation.

## Fund raising and finance

Towards the end of the WARA campaign, John Clare of *The Times* described WARA's efforts as being 'almost disturbingly professional' in a report headed 'Foulness backed by big money lobby'.[38] This provoked angry replies from WARA supporters who resented the implication, as they saw it, that North Buckinghamshire was run by a sinister body of wealthy and politically powerful oligarchs. Throughout its life the WARA Executive seemed self-conscious about its undoubted ability to raise money locally and made sure that properly audited accounts of expenditure were readily available. This is rather fortunate as the main criticism at the end of WARA's campaign was that it was a professional public relations exercise. We shall discuss this criticism later, but suffice it to say at this stage that publicity and public relations eventually accounted for only 16 per cent of the total WARA expenditure over two years. The vast bulk of expenditure went on legal and professional fees, totalling £43,713 and representing 76 per cent of the total expenditure. A full account of WARA's expenditure is reproduced below:

Receipts and payments account for the period from 10 March 1969 to 21 June 1971.

| | £ | £ |
|---|---|---|
| RECEIPTS | | |
| Donations and proceeds of fund-raising activities .. .. .. .. .. .. .. .. .. | | 57,415 |
| Building society interest .. .. .. .. .. .. | | 395 |
| | | 57,810 |
| *Less* | | |
| PAYMENTS | | |
| *Legal and professional* | | |
| Solicitors (Lovell, White & King) .. .. .. .. | 7,147 | |
| Counsel (Niall MacDermot, QC, and Philip Otton) .. .. .. .. .. .. .. .. | 15,475 | |
| Noise experts (W. F. Atkins & Partners) .. .. .. | 10,125 | |
| Air traffic control experts (Alan Stratford & Associates) .. .. .. .. .. .. .. | 3,679 | |
| Planning experts (Triad) .. .. .. .. .. .. | 6,635 | |
| Report and transcripts of evidence .. .. .. .. | 652 | |
| (76 per cent of total expenditure) .. .. .. .. | 43,713 | |
| *Publicity and public relations* | | |
| Publications .. .. .. .. .. .. .. .. .. | 4,750 | |
| Leaflets and Publications .. .. .. .. .. .. | 1,984 | |
| Signs and posters .. .. .. .. .. .. .. .. | 1,532 | |
| Press advertising .. .. .. .. .. .. .. .. | 814 | |
| Cost of meetings .. .. .. .. .. .. .. .. | 264 | |
| (16·3 per cent of total expenditure) .. .. .. | 9,344 | |
| *Administration* | | |
| Staff costs .. .. .. .. .. .. .. .. .. | 3,232 | |
| Postages, stationery and sundries .. .. .. .. | 951 | |
| (7·3 per cent of total expenditure).. .. .. .. | 4,183 | |
| | | 57,240 |
| FUND at 21 June 1971 .. .. .. .. .. .. | | 570 |

Of equal interest is the way in which this huge sum of money was raised. We have noted earlier the speed with which WARA managed to accumulate funds or promises of funds and the confidence which the leaders of the Association exhibited in planning their strategy at the outset. But this does not mean that the Association had no financial difficulties throughout its life. At the Executive Committee meeting in June 1969, Derek Lawson expressed concern at the financial situation in view of the likely cost of the Aylesbury hearing, though Fennell commented that people worth £5,000,000 always took a long time parting with £5,000 and that was why they were worth £5,000,000![39]

The initial response to the WARA appeal had been most encouraging (though £5,000 of the promised money never materialized), but after the Stage II Aylesbury hearing the Executive had to make an important decision about whether to continue the legal and technical representations. The matter was pressing because Bedfordshire CC still wanted Niall MacDermot to represent them at the Stage V hearing in London. After some discussion it was agreed that MacDermot should continue to represent the Association through to Stage V and that a minimum of a further £20,000 should be raised to defray the costs of this. At Fennell's suggestion it was further agreed that future fund raising should be divided into two parts: the individual approach and the group activities, organized by the various village committees under the Grass Roots Committee. Geoffrey Ginn estimated that a further £10,000 could be raised by the Grass Roots Committee, thus leaving £10,000 to be raised by the individual approach. A small sub-committee was formed, under Evelyn de Rothschild, to organize the raising of additional funds from individuals. This move was delayed, at Rothschild's suggestion, until the first six months' accounts had been audited and widely distributed, so that those approached a second time would know what had happened to earlier donations. Although the sub-committee met only once and had delayed its activities, the individual approach was remarkably successful. Accounts dated 8 March 1971 showed that during the whole two-year period £25,852 was received by WARA in *individual* donations. Of this sum over £10,000 was subscribed by twenty-one donors (i.e. 41·4 per cent of individual donations and 20 per cent of all receipts). The remaining £15,000 was donated by 2,520 individuals. The personal standing of members of the sub-committee, containing as it did Lady Hartwell, Major Phillip Duncombe (Regional Secretary of the CLA Bedfordshire, Buckinghamshire, Hertfordshire, Middlesex, and Oxford Branches), Bill Manning, Alec Miscampbell, A. Hugh-Smith and Vivian Street, no doubt facilitated the collection of funds. In addition, a donation of £1,000 was received from the National Trust and £1,000 from the Vale of Aylesbury and Brill Area Branch of the CPRE (the latter contribution coming from an anonymous private donor and earmarked for public relations). It is difficult to estimate the extent to which strictly commercial interests contributed to WARA's funds but it appears that such contributions were relatively unimportant. The Executive Committee considered an approach to local firms through a fund-raising lunch but this was never held. The donation of £500 from the Deanshanger Oxide Works was the only substantial industrial contribution, though there is some evidence to suggest that Bedfordshire Airport Resistance Association (BARA) was more fortunate and obtained considerable help from Unilever.[40] There is certainly no evidence to suggest that the commercial interests concerned in promoting Foulness as a possible site (mainly the Thames Estuary Development Company, but also the Thames Aeroport Group) in any way contributed to the Association financially. In a BBC-2 programme on the Roskill Commission it was vaguely suggested that TEDCO had contributed to WARA's funds, but if this was true then Desmond Fennell was unaware of it, for he wrote to TEDCO saying that he believed the suggestions to be untrue and asking for confirmation.[41] It seems unlikely that Fennell would have been unaware of any financial assistance from TEDCO, though ultimately one cannot assert that there was no such assistance since WARA did receive an anonymous donation of £1,000, the source of which was never discovered. However, this could equally have come from a prominent politician, or other public figure, or indeed a shy nonentity.

The remainder of WARA's finances (slightly over 50 per cent) came from its own fund-raising events, the efforts of its individual committees and from supporting associations. These events ranged from local whist drives, fêtes and jumble sales (for example a jumble sale at Stewkley raised £68, and a coffee morning at Aston Abbotts raised £35 towards its eventual contribution of £1,000 from a population of 270) to large fund-raising events organized by WARA itself. Of particular note in the latter category was the 'At Home' organized by Lady Hartwell at Oving House. Sir John Betjeman agreed to speak on 'The threat to England and its heritage', and the event produced a profit of over £1,000 for WARA. Similarly an Open Day, followed by a supper at the Hanstead Stud, produced a further £1,000, a Christmas Fayre produced over £2,000, a sponsored walk raised £2,700 and a Strawberry Fayre produced £600. It would be impossible to list here all the events and fund-raising ideas used by WARA but almost every conventional method of raising money was used, plus quite a number of novel ones. Some of the more ingenious included ladies' football, autocross, a scrap-metal drive, an Easter bonnet contest, a wig party, and even an air pageant!

The WARA newsletter, apart from providing information on the progress of the campaign, devoted considerable space to reporting the fund-raising efforts held in the various villages and occasionally singled out particular individuals (e.g. a lady who collected over £70, door-to-door) or particular villages (e.g. Aston Abbotts and Stewkley) for special praise. In doing so, it no doubt encouraged the rank and file to redouble their efforts and kept up morale.

Although many of the items listed in the newsletter may have seemed trivial, e.g. £7 raised at a film evening, even minor fund-raising efforts were essential to WARA's task for, despite large individual subscriptions, WARA sometimes had to delay payment of large and important accounts until more ready cash could be found. This was particularly so during the approach to the Stage V hearing in London (which WARA had recognized as the most important stage of the Roskill hearings), and Lovell, White & King wrote to Fennell about the 'vexed question of costs in connexion with the forthcoming Stage V Inquiry'. The main problem, from the solicitors' point of view, was that they would technically be liable for Counsel's fees, estimated to be in the region of 14,000 guineas. Basically they were unwilling to brief Counsel for Stage V unless they had some surety from WARA, and thus Fennell was informed that Lovell, White & King would not deliver the brief 'without a fully realistic payment on account beforehand'. In fact they wanted WARA to deposit £20,000 in advance of the Stage V hearing, but in the event Rothschild as Treasurer agreed to the payment of £8,000 in advance as WARA had that sum on deposit with the Anglia Building Society. He wrote to Fennell in May 1970 saying that he did not think that more money should be set aside as 'we are respectable people and I hope that our solicitors think we are'.[42]

WARA was by any standards a wealthy pressure group; certainly by those of environmental pressure groups. One supporter, who sat on a number of committees concerned with aspects of the environment, wrote to Fennell saying that he was staggered that WARA had spent so much money. However, though finance was normally subsidiary to tactical considerations and rarely was a course of action rejected solely on financial grounds, money matters occupied a good deal of time and thought. Financial problems never really became serious, not only because of Rothschild's skill as a fund-raiser, but also because of his resources and ability at financial management. He provided both ideas and practical help; he established that local farmers could deduct contributions to WARA from taxable profits providing that they were supporting the

Association in order to avoid damage to their business or loss of agricultural land, and he sent one of his employees, the Ascott Estate Accountant, to the WARA office twice weekly to help with the accounts. Also, as the large accounts arrived in June 1970, Rothschild arranged overdraft facilities of £5,000 from the local branch of the National Westminster Bank. Although they were never pressed unduly hard to settle accounts quickly, no doubt due to the standing of the Association's Officers, Rothschild delayed the payment of some accounts so that receipts and payments did not get too far out of step. Throughout the campaign he kept tight control over the finances and insisted that no significant account was paid without his prior approval.[43]

## Co-ordination with other organizations

WARA, of course, was not the only organization concerned in the airport controversy. In addition to other anti-airport groups and amenity societies, local authorities were involved, as were the consortia anxious to develop a coastal site. From WARA's point of view, perhaps the other resistance organizations directly concerned with the siting of the airport provided the most difficult problem of liaison.

A common feature of environmental problems of this nature is that the battle for the environment can be in danger of being weakened by serious conflict between groups each trying to push the proposed development into the other's territory. Alternatively, the absence of real organized pressure in one area may mean that any proposed development may be pushed in that direction. As we shall see in our concluding section there are at least prima facie grounds for suggesting that this was indeed the situation regarding the airport when the Government eventually had to reach a decision. Almost from the outset the opponents of the three inland sites were in favour of siting the airport at Foulness and they could all see the advantage of a combined resistance movement against *any* inland site. Not until the final stages of the whole airport controversy, however, did anything like a joint anti-inland front emerge. This is not to suggest that a joint approach was not attempted earlier. The Bedfordshire Airport Resistance Association contacted WARA in 1969 suggesting a joint approach (possibly in the form of a joint press statement advocating Foulness) but the WARA Executive saw 'considerable difficulties' in the idea, not the least of which was that an association of this sort 'was bound to be fairly parochial'.[44] This view was still held by WARA in April 1970 when, in response to a newspaper report of a meeting at which it was asked why WARA had not joined with other resistance organizations, Pargeter wrote that if an inland site was chosen, then the clear duty of WARA to its members 'is to oppose the siting at Wing even if the result may be to site it at one of the other two inland sites'.[45] In fact WARA could ill afford to compromise with other groups, since they recognized that it was always a possibility that Foulness might be rejected by Roskill. Indeed the Clerk to the Buckinghamshire County Council had written to Pargeter in April 1969 suggesting that the more one looked at the problem, the more it looked as though on pure cost and economic factors, Wing might well come out on top of the resulting balance sheet. Pargeter's views were reiterated in July 1970 by Fennell after WARA's planning consultant, in evidence to the Stage V London inquiry, suggested that if Foulness was not selected, then Thurleigh was the best inland site as it had the greatest potential as a major airport conurbation. Fennell stated that WARA had to be prepared for the worst and had to protect the interests of the people they represented and who had paid for the case which WARA was presenting.[46]

In a letter to BARA designed to improve the strained relationship which this development produced, Fennell explained away the fact that WARA had suggested Thurleigh on the ground that Mr Justice Roskill had asked Counsel for their client's views about the respective merits of each site. He assured BARA that whatever the position during the Commission's inquiry, there was no question of WARA advocating anything other than Foulness. He thus offered an olive branch on behalf of WARA in the hope that the inland sites could 'once again join forces'.[47] It is unlikely that WARA had the intention of maintaining this united front in all circumstances; it was always at the back of their minds that they might have to break with Thurleigh and pursue an independent policy.[48] However, this approach received a warm response from BARA which suggested that they should join forces in the forthcoming Parliamentary battle. Shortly afterwards, Fennell wrote to other resistance organizations (Nuthampstead Preservation Association, Gatwick Anti-Noise Executive, Kew Association for the Control of Aircraft Noise, and Luton and District Association for the Control of Aircraft Noise, LADACAN) who agreed to support the Parliamentary moves. The role of the CPRE in this joint approach will be discussed later, as will the importance of the Parliamentary battle against the airport; but, by December 1970, John Pargeter was able to write to the Chairman of the Gatwick Anti-Noise Executive informing him that the resistance associations of the three inland sites had agreed to prepare a joint initiative. They had decided to support a backbench Parliamentary Committee, whose object was to urge the Government 'not to permit any inland site for the Third London Airport or *any further extension of existing inland sites*'. It is interesting to note the broad outlook which WARA had adopted, as by then (late 1970) it was becoming clear that Wing was the front runner as far as Roskill was concerned. Clearly it was in WARA's interest, in view of this weakening of its overall strategic position, to gather as much support as possible from the other resistance organizations. It was for this reason that Pargeter had stressed that 'the problem was now greater than just the choice of the site for the Third London Airport' and that the problem was 'basically whether international airports of this size are to be based on inland or on coastal sites'.[49] The main problem which WARA faced at this stage was not so much with the organizations at the short-listed sites but rather with those at the existing airports, who feared that building the new airport at Foulness would give rise to a greater increase in air traffic at their own airports than if Cublington were to be chosen.

We have noted earlier that, apart from contact with other resistance associations, WARA had good links with a whole range of existing voluntary societies in the area. This was particularly true for some of the amenity societies such as the CPRE. In fact WARA and the CPRE (both at the national and local level) had an identity of interest in that the CPRE was totally opposed to any inland site for the airport and retained Counsel (Richard Body, MP) to appear for each of the county branches in turn at the four sites affected, and also before the Roskill Commission at the Stage V hearing in London. There was also activity at the branch level: at a very early stage in the life of SARA contact was made with R. G. Alexander, the Secretary of the CPRE Buckinghamshire branch and although events moved at such a pace that the local CPRE sometimes 'ceased to hear the band', at no time was 'formation' really broken.[50] Thus, while the CPRE had its own role to play, the Buckinghamshire branch in particular decided to support and supplement WARA's efforts. When it was learnt, in December 1970, that Roskill had chosen Wing as the best site, the Chairmen of the Buckinghamshire, Northamptonshire and Oxfordshire branches of the CPRE issued a joint statement supporting WARA (and the County Councils) in their efforts to persuade the Government to reject the recommendations.[51]

Of great importance to WARA was the CPRE's role as a *national* organization and particularly that of its respected Chairman, Lord Molson. The intervention of Lord Molson on a personal level was regarded by some members of WARA as a high spot in the whole campaign as it helped to transform what was still essentially a local issue into one of national significance. Molson's views received considerable national publicity when, together with Duncan

Sandys, MP (President of the Civic Trust), Peter Shepheard (President of the Royal Institute of British Architects), Walter Bor (President of the Town Planning Institute), and J. St Bodfan Gruffydd (President of the Institute of Landscape Architects), he wrote to *The Times* on 14 January 1971 (i.e. after the Roskill decision had been announced) expressing 'profound concern' that Wing was proposed as the site for the airport. The signatories held the view that the Commission had accorded insufficient priority to the protection of the environment and important planning considerations, and had paid too little attention to the arguments advanced by the Countryside Commission in favour of Foulness. Molson supplemented his intervention in two important ways. First, he took the initiative in organizing a meeting of all anti-inland airport groups concerned in the controversy and with skilful chairmanship managed to persuade them to agree on a joint statement suggesting Foulness as the most suitable site.[52] This agreement was no mean achievement for, as we have noted, each group tended to be parochial in its main aims. This was especially true of LADACAN, but even it agreed not to break ranks. Second, Molson became the leader of the opposition to Roskill in the Lords and played a crucial role in ensuring that the Lords eventually came out strongly against the Report.

The other major interests with which WARA had to co-operate were the local authorities. Since they were all opposed to an inland site, WARA was saved one major exercise in lobbying, but this did not mean that there were no problems. The main bone of contention lay in the role which WARA was to play, particularly in relation to the Roskill Commission. We have seen earlier that SARA experienced some difficulty in first establishing credibility with the local authorities. However, the Councils soon perceived that an association like SARA could play a useful part in the campaign and, when it became clear that the battlefield would be Wing and not Silverstone, they encouraged Fennell to continue the fight further south. The Chairman of Buckinghamshire Planning Committee, Ralph Verney, rang Fennell to discuss the new situation. They agreed that whatever protest was made by the statutory bodies it had to be reinforced by individual protest and that WARA was the most suitable organization to accomplish this. Thus the Association had the blessing of the County Council, via Ralph Verney, from an early stage. However, WARA became particularly concerned lest its role be restricted to the initial brief of being merely the vehicle for individual protest. A meeting with the County Council at Aylesbury in March 1969 did not allay WARA's fears, and its leaders left the meeting dissatisfied and convinced they had to go it alone. Their personal involvement was so great that they were not content to leave all the technical representations to the County Council. However, shortly afterwards a second meeting was held, this time at the premises of Nathaniel Lichfield, one of the Councils' consultants, to enable the two sets of experts to meet and exchange information. This meeting proved more satisfactory and the Council finally acknowledged the ability and stature of WARA's experts. That it was accepted that WARA should play a major part in presenting the technical case did not lessen the need to reinforce this with a 'spontaneous expression of organized and informed private indignation', as Verney put it.[53] He saw WARA as representing the area's grass roots – 'the Yeomen of Cublington rising up in their wrath'. Once relative roles had been established, liaison with the Councils became very close; Bedfordshire CC even used Niall MacDermot, WARA's Counsel at the Aylesbury hearing. The relationship with the parishes was also close, for WARA took many of the Buckinghamshire and Bedfordshire Parish Councils under its umbrella since the appropriate Association of Parish Councils did not have the facilities to organize the parishes' representations to Roskill. It was fortunate for WARA that its case was in no way weakened by the attitude of the local authorities. Had the latter been in favour of the airport, WARA's task would have been considerably more difficult. As it was, the Buckinghamshire County Council was even prepared to donate £1,000 to WARA at the end of the campaign to help pay outstanding debts. It was equally fortunate that other major bodies affected by the airport proposals shared the opposition of WARA and the Councils. The Governors of the proposed Cranfield Institute of Technology (whose Professor of Aircraft Design, D. Keith-Lucas, was a member of the Roskill Commission) decided to associate themselves with WARA's protest and were prepared, if asked, to make a financial contribution.[54] Also, Lord Campbell of Eskan (Chairman of the Milton Keynes Development Corporation) stated at the Aylesbury hearing that building the airport at Wing would mean that the plans for the new city would more or less have to be torn up. Lovell, White & King held the view that this statement was of great assistance to their case.[55]

## General strategy

The general strategy adopted by WARA was of course largely dictated by the procedure of the Roskill Commission and consequently there were relatively distinct stages in WARA's life which broadly corresponded to the various stages of the Roskill inquiry. We have already noted that the first aims were to establish credibility, to enrol membership and to raise money to pay for the expert representation before the Commission. As the Stage II hearing approached, the campaign began to develop on two levels. The legal and technical battle was organized mostly from London by the so-called London Committee, an informal group which had its origins in a suggestion from Evelyn de Rothschild during the Maxwell row.[56] This somewhat detached process which involved few people and gobbled enormous sums of money, continued in London while the threatened areas organized local activities, the main aims of which were to raise money and to give ordinary members of the population an opportunity to participate. This was an important aspect of the campaign for two reasons. First, by participating in events, people could feel that they were contributing to the campaign directly, rather than merely digging into their pockets to finance a technical wrangle which they could not understand. This activity also improved the communication situation which was frequently under attack; WARA was often accused of being cut off from its membership, even by some of its leading supporters.[57] Second, the involvement of the population at large helped at one and the same time to dispel the 'Roseberry' image and to demonstrate to the Government the intensity of local feeling.

The local activities were too numerous to be illustrated in detail here. As far as participation was concerned, the high point was 'Operation Green Man', which began as a somewhat extremist proposal that all the roads in the area should be blocked in order to demonstrate the strength of the opposition to the airport, and ended by being precisely the opposite – an open invitation to the Nation to visit the area and to take tea with the local people.[58] However, while it was an essential precondition of success that WARA had to show that the people of Wing *were* opposed to the airport, it would have been foolish to have concentrated entirely on the articulation of mass opposition to the exclusion of more technical opposition. Thus WARA had to challenge the planners on their own ground by demonstrating that, on objective criteria, Wing was *not* a more suitable site than Foulness. In retrospect, WARA can justifiably claim that it was successful both in demonstrating mass popular support and in presenting a reasoned and responsible technical case to the Commission. For example, the Inspector at the Aylesbury hearing commended the way in which WARA had presented its case, and indeed Mr Justice Roskill praised the closing speech made by Niall MacDermot at the Stage V hearing. He observed that if he were to list the half dozen most valuable contributions which the Commission had had 'the contribution of WARA, would be among them'. It was his view that the 60,000 WARA members could feel that nothing they wished to say had gone unsaid in the seventy-three days of the London hearing, as well as in the proceedings at Aylesbury.[59]

Having concluded its submissions to the Roskill Commission, WARA had to consider what further strategy should be adopted, and in particular what should be done when the Report was presented. The Association's activities focused on four main topics: technical evaluation of the Roskill Report, local activities, the role of public relations and the nature of Parliamentary opposition to Roskill.

In order to be in a position to provide effective refutation of Roskill's arguments when the Report was published, Fennell approached a number of eminent people who supported WARA and had expertise which might be useful. The group, known collectively as the 'Think Tank', was given the very difficult task of reading the Report and producing detailed criticisms within forty-eight hours. Although not wholly successful in attaining this objective, the Think Tank did contribute a number of ideas which were embodied in WARA's critique. So that the Association should have something to say on the day of publication itself, a document was prepared in advance offering general criticisms of the Commission's approach and of those arguments which could be inferred from the hearings. This critique was unveiled at the first press conference, arranged at less than twenty-four hours notice by WARA's public relations firm, Burson-Marsteller. It was fortunate that on 17 December 1970, Lord Strabolgi, working for Burson-Marsteller, had heard that a summary of the Report was to be presented the next day, thus enabling WARA to stage a press conference, attended by the national press, radio and television, within five hours of the publication of Roskill's conclusions. A more detailed criticism of Roskill was presented at the second press conference, in January 1971, three days after the full Report had been published.

Of course not all the activity was confined to London. In the ordinary course of events a number of functions had been arranged for the weekend before Christmas, including a Christmas Fayre, a Carol Service, and 'It's a Knockout'. The addition of activities specially organized for the publication of Roskill, such as the lighting of the beacon bonfires on over thirty hilltops in the area, accompanied by the tolling of church bells in 'muffled sympathy', gave to some observers the impression of a slick professional organization at work. Though WARA's organization was undoubtedly professional, the overall effect on this occasion was more fortuitous than designed. In order to drive home WARA's case, a series of events were organized after Christmas including a motorcade of vehicles, the huge rally at the Wing Equestrian Centre, a protest at Westminster by 500 women, the open day held by seventeen villages, and the Great Environmental Pancake Race.

Despite its undoubted success, both in its case to Roskill and also in demonstrating its mass support, the WARA Executive had a long internal debate concerning a further strategy for its campaign. This was the possible creation of a favourable climate of opinion, both in informed circles and with the public at large, by means of public-relations techniques. Something of a myth has developed about the nature of this aspect of WARA's campaign. For example, it has been suggested that the Cublington campaign proves conclusively that public relations do work.[60] It has been claimed that WARA's campaign was '*directed* by a clever team of professional public relations men working for a high-powered American-based company'.[61] Similar stress was laid on the role of the P-R firm in the controversial article by John Clare, referred to earlier. Clare quotes a principal of the firm as describing WARA as a 'body of complete amateurs who knew nothing at all about dealing with the press'. If accurately reported, this was a singularly inappropriate comment in view of the fact that the wife of the Chairman and Editor-in-Chief of the *Daily Telegraph* was a member of the Executive Committee, and that the Press Officer was a successful free-lance journalist, who subsequently joined ITN. Less surprisingly perhaps, the P-R industry itself has seen the WARA campaign as proof of the efficiency of their work. In an article entitled 'How public relations used the "human approach" to turn the choice of London's Third Airport from Cublington to Foulness', in the P-R journal *Campaign*, the leader of the Action Committee Against Foulness Airport is quoted as saying that WARA's ready access to Government, MPs, Press, TV and Radio had been the most depressing feature of the whole affair.[62] However, there is serious doubt about the veracity of some of these claims, and in the same article even Claude Simmonds (Chairman of the P-R firm concerned) is reported as saying that the agency's fee was 'very nominal indeed' towards the end of the campaign. He had stuck with it because 'I'm very much a countryman. It's true that Desmond Fennell's wife is my niece, but this was never any part of the decision to retain me. The Association could have handled the whole campaign itself, and it would have won, but it wouldn't have done it half so well.'

The suggestion that a professional P-R agency should be engaged to act for WARA was canvassed as early as June 1969, although a formal decision to do so was not reached until WARA's campaign was well advanced. The question was investigated by a small subcommittee which drew up a short-list of P-R consultants and held preliminary discussions with CS Services, a British subsidiary of an international P-R group. By October 1969 the sub-committee was in favour of employing CS Services on the basis of £500 per month starting in January, after a preliminary investigation at £1,000. The Executive, interestingly enough, suggested that Niall MacDermot's views should be sought, and left the ultimate decision to the London Committee. Two factors weighed in favour of employing a professional firm. In the first place, so much of the budget had been spent on the lawyers and technical experts that it was difficult to make a case for not spending a small sum on P-R, and second, money had been offered to WARA expressly for P-R activities. Eventually CS Services were instructed to carry out a preliminary investigation and this was completed early in 1970. A full discussion of the report took place at the Executive meeting of 5 April, and it was agreed that a campaign should be conducted by CS Services for three months at £500 per month, and should concentrate on furthering the choice of Foulness and creating public awareness of the folly of choosing an inland site – particularly Wing.[63]

At the end of the first three months of the campaign the Executive evaluated what had been achieved, and it decided to continue at the £500 rate for only one further month. Thereafter, CS Services, who by then had metamorphosed to Burson-Marsteller Ltd, would be retained at £100 a month until the publication of the Roskill Report. It is not easy to evaluate what was due directly to Burson-Marsteller's activity and what WARA would have achieved on its own. For their part, Burson-Marsteller claimed a long list of successes, including responsibility for the good coverage in *The Times* and the *Daily Telegraph* of Professor Hancock's evidence to the Commission. They certainly arranged a meeting with Andrew Wilson of the *Observer*, concerning a possible article on all the groups opposed to inland sites. Press statements were also issued on WARA's behalf and contacts were made with many different sections of the media. In particular an interview was arranged for Fennell on the *Today* programme, and a camera team from *Nationwide* covered the 'At Home' at Oving, where Fennell and John Betjeman were interviewed (although, because of the way in which it was edited, WARA considered this programme to have been of little help). It is doubtful that WARA could have undertaken all this itself.

In October Burson-Marsteller produced a plan for the next, most important, phase of the campaign. This proposed a national press conference when the Roskill Report was published, with a follow-up conference when more considered arguments could be put forward; until then, they would remain on the £100 per month retainer. However, seeds of doubt had been sown among committee members and a considerable debate ensued about the role of public relations, after which it was agreed that the Public Relations Sub-

Committee should meet one of Burson-Marsteller's new directors, a former Director of Publicity at Conservative Central Office. At the Executive Committee meeting on 6 December 1970 it was reported that the sub-committee was still split on the problem, and at the end of the Committee discussion it was resolved to continue with Burson-Marsteller, but that the firm should be asked to produce detailed plans and would be supervised by Messrs Lonie and Lovejoy.[64] Burson-Marsteller was perhaps unfortunate in that these discussions coincided with a period of financial difficulty for WARA which fact naturally inclined the Association to rely on its own resources. In fairness to the agency it must be said that, when they realized WARA's financial position, they agreed to write off the Association's outstanding balance of £344 in December 1970, and to continue on the £100 per month retainer with only a possibility of reverting to £500 after the publication of the Roskill Report. During this period Burson-Marsteller were quite active on WARA's behalf. They kept the Committee informed of the build-up to the House of Lords debate on the Roskill Report; they arranged for the printing and distribution of WARA's criticism of the Report and arranged the press conferences at which these were unveiled; they maintained a flow of information to the responsive journals and to the lobby, regional planning and amenity correspondents; they managed to place some of the letters written by WARA's letter-writing panel on radio programmes like *Listening Post*, and they provided some highly valuable and accurate information from within Whitehall.

Thus, the claim that the WARA campaign was *directed* by Burson-Marsteller is absolute nonsense. The agency never reverted to its £500 per month fee and never really played the role in WARA's efforts which many observers have implied. That the Burson-Marsteller account totalled little more than £3,500 is itself a good indicator of the firm's role in the whole affair. Even when WARA's expenditure on all forms of publicity and P-R is taken into account – and this includes the cost of publications, leaflets, signs and posters, press advertising and the cost of meetings – it is still under £10,000 and represents only 16 per cent of the Association's total expenditure. As Desmond Fennell commented afterwards, in response to the suggestion that their campaign was a well-oiled victory for the P-R industry: if people really knew the true facts 'the idea would have been so laughable as to have never been considered'.[65] Where Burson-Marsteller contributed most was in the organization of the press conferences (at very short notice), the contacts with various journalists (especially at the BBC), and information gleaned from their impeccable contacts. Part of the reason for the unease that some (though by no means all) members of the WARA Executive felt over the use of a P-R firm was that, in many ways, WARA did not need this kind of professional advice; the airport issue was *inherently* of interest to the mass media and hence, eventually, to Parliament. Furthermore it so happened that many of WARA's activities were 'good television' (this was especially true after the publication of the Roskill Report). They were almost certain to attract wide press and television coverage whether or not professional P-R men were employed. Further, WARA's own Press Officer, John Flewin, was extremely able and, as a free-lance reporter for the national newspapers who also worked for the radio programme *South-East,* was able to secure considerable publicity. Thus the P-R industry had less by way of expertise to offer the Association than many people think, and WARA was never *dependent* on Burson-Marsteller. Indeed, the latter's contribution was mainly one of high-lighting certain aspects of the campaign by providing London-based facilities not readily available to WARA. Other pressure groups, not so skilled or so well-connected, may find that their need of the industry's services is more acute.

It is ironic that the WARA campaign should have been criticized as a victory for professional P-R techniques when in fact relatively little of the Executive's energies were devoted to that aspect of the campaign. Greater effort, both before the publication of the Roskill Report and afterwards, was put into 'behind the scenes' work than into creating a particularly favourable public image. Publicity and the creation of a 'human image' were important, but the Executive was always conscious that the manner in which the Roskill Report would be handled could prove crucial in the end. Consequently considerable effort was put into trying to establish exactly how the Government would handle the Report. WARA's activity in this direction is a perfect example of what Professor Finer has termed 'advance intelligence'.[66] He has described the necessary conditions for the success of a pressure group under the general heading of the 'technical efficiency of the lobby'. These conditions are (1) advance intelligence, (2) established access to friendly MPs and (3) facilities for briefing. As we shall see, WARA met all three conditions and if, as John Clare argued in *The Times,* WARA was 'disturbingly professional', then it was in meeting these three preconditions of success rather than in being a well-oiled P-R machine.

Apart from the financial resources so necessary to present a reasoned case to Roskill, WARA possessed a second and more important advantage. It had excellent contacts with people familiar with, or part of, the Westminster/Whitehall circuit. In short, the WARA leaders and advisers were part of what Ely Devons has termed the 'Inner Circle of journalists, academics, men of affairs, civil servants and public commentators'.[67] As Richard Rose suggests, the majority of persons in national political roles are drawn from a fairly homogeneous social class, facilitating horizontal communication and impeding vertical communication.[68] It so happens that most members of the WARA Executive were drawn from this social stratum and hence were able, to an unusual degree, to 'sound out' opinion in Whitehall and Westminster on the way in which events were likely to run.[69] While Fennell was, of course, interested in the Government's attitude to the Roskill Commission, and its possible recommendations, he was particularly anxious to establish the procedure which would be followed when the Report was submitted. It was to WARA's great advantage that some of its supporters had family contacts within the civil service departments directly concerned; for example one member wrote to Fennell saying that he had asked a member of his family at the Board of Trade (as it then was) 'to do some spying'. As early as March 1970, Fennell received a note from a contact in the Foreign Office suggesting that consistency with the objectives of European policy would determine the choice of site and that a Thames Estuary location would fit naturally into the industrial and social pattern which would develop in that area over the next decade. Eventually Fennell wrote to several people, either in Whitehall or who had good connexions there, asking if they could discover whether or not the Government would reach a decision on Roskill's recommendations before publication of the Report, or whether it would allow a period between publication and the final decision. The replies were, not surprisingly, somewhat contradictory. One contact, of considerable status and experience, suggested that the Government would reach a decision before publication and that to allow half-formed opinion loose on matters before making a policy statement would not be in line with Edward Heath's 'we're in command' approach. Fortunately for WARA this was *not* the correct prediction and indeed another contact with friends in central government departments experienced at handling reports of Royal Commissions, held the view that the Government *would* allow a pause between publication of the Roskill Report and a policy statement. This would be in line, he argued, with the Government's 'we are not going to be rushed' approach. Yet another correspondent wrote to Fennell confirming this view, although warning him that Wing was likely to be top of Roskill's list of sites. He therefore advised that WARA should make an all-out effort to discredit the findings of the Commission, should mount the most intense lobby of MPs and Ministers, and should create a public outcry designed to convince the Government that the acceptance of the Report was *not* the line of least resistance. The writer concluded by saying that he was having lunch with the Chancellor of the Exchequer in November, and also hoped to be entertaining the Prime Minister at a future date and so would have a good opportunity to sound them out on post-Roskill procedure. Fennell was further advised to ensure that the Government did not announce a decision prior to or on the publication of Roskill; WARA should encourage its Parliamentary supporters to put down Questions in order to obtain an undertaking from the Minister (Frederick Corfield, Board of Trade) that the House of Commons would be given a full opportunity to debate the issue before a decision was made by the Government. Thus on 22 July 1970, Mr Goronwy Roberts (Lab., Caernarvon) asked the Minister to confirm 'that he will not necessarily be tied to the Roskill proposals' and to confirm that the House would be given adequate opportunity of debating any decision that the Government might recommend. On 23 November, Stephen Hastings (Cons., Mid-Bedfordshire) asked for a similar assurance that the Report would be made public at the same time that it was given to the Government. A week later W. Benyon (Cons., Buckingham) asked the Minister when he expected the Report and what procedure would be followed thereafter and whether the Government would allow time for debate in the country and in the House before making up their minds on Roskill's recommendations. Benyon was supported by Nigel Spearing (Lab., Acton) and the Minister did give an assurance that he was conscious of the need for the matter to be debated in the House.[70] The matter was clinched, however, in a rather unusual way. On 12 January 1971, William Whitelaw, the Leader of the House of Commons, took part in the radio programme *It's your line* and, in answer to a telephone call from Desmond Fennell, gave an assurance that Parliament would have every opportunity to debate the Roskill Report before the Government announced its decision. Shortly afterwards he sent his PPS to Benyon suggesting that he should ask a Parliamentary Question in order that the debate might be announced in the reply.

The net result of the Association's excellent contacts within the administration, both through its supporters and through Burson-Marsteller, was that it managed to gain sufficiently reliable advance intelligence of the likely procedure for handling the Roskill Report that it was able to plan what was to prove the most crucial phase of the whole campaign, namely the Parliamentary battle. Since the Government allowed a considerable period for discussion before announcing its decision on the airport, the Parliamentary contacts proved to be vitally important. Here again, WARA was able to fulfil an important precondition of success.

By June 1970, it had become clear to WARA that the selection of

Cublington was a possibility and that therefore a change of strategy was required. Having pulled out all the technical stops they could, it was clear that they then had to become more politically orientated. On 9 July a meeting was held between local MPs (W. Benyon, D. Madel and T. Raison) and representatives of WARA and of Burson-Marsteller.[71] The next day a press statement was issued announcing the formation of an All-Party Committee with the 'express object of co-ordinating opposition to the choice of any of the proposed inland sites at Wing, Thurleigh or Nuthampstead'. The statement expressed grave doubts about the cost-benefit methodology employed by Roskill and argued that present conditions and future trends in this country dictated that the proper place for a large international airport should be Foulness. The Committee announced its intention to keep in touch with the relevant County Councils and resistance associations. Stephen Hastings, who had formed an informal Parliamentary group to press the case against an inland site as early as April 1969,[72] became the Chairman of the new Committee and William Benyon, who had defeated Robert Maxwell at Buckingham, became the Secretary. The other founder members were the Rt. Hon. Sir David Renton, QC (Cons., Huntingdonshire), James Allason (Cons., Hemel Hempstead), Peter Fry (Cons., Wellingborough), David Madel (Cons., Bedfordshire, South), Timothy Raison (Cons., Aylesbury) and Trevor Skeet (Cons., Bedford). In order to emphasize the all-party nature of the Committee, Mrs Shirley Williams (Lab., Hitchin) was later brought in as Vice-Chairman.

The All-Party Committee used three main tactics in waging its campaign. First, it organized a series of meetings and visits aimed at informing and persuading MPs. It is quite common for all-party committees to be addressed by 'outsiders', and among the speakers invited was Desmond Fennell who, on 2 December, gave the Committee a résumé of WARA's case against the airport. In his speech Fennell stressed that, whatever Roskill recommended, it would be the House of Commons that would eventually decide where the airport would be sited. He particularly emphasized the restrictive terms of reference which Roskill had been given, and that the fundamental point at issue was, 'What sort of world do we want to live in?' His speech was extremely well received by the MPs, and WARA received a second invitation to meet the Committee on 9 February. On this occasion Fennell was accompanied by WARA's expert consultants so that the MPs could be fully briefed on the Association's technical case. In addition to hearing WARA's arguments, the Committee met representatives from other groups, though it is clear that it was WARA who made the most impact.

Second, the All-Party Committee set as its main task the marshalling of massive backbench opposition to the siting of an airport on any inland site. The operation was mounted by means of an 'Early Day Motion'[73] which opposed the selection of any inland site and advocated Foulness or any other suitable coastal site. Considerable care was taken in drafting the Motion in order to maintain the unity of the various outside groups. It was important from the All-Party Committee's point of view, to avoid breaking the sometimes fragile unity amongst the resistance associations. Equal care was taken in enlisting Parliamentary support for the Motion and the Committee decided to aim at attracting at least 150 signatures. In fact Parliamentary supporters were 'lined up' in advance so that once the Roskill findings were published, they could immediately demonstrate, if need be, that a large proportion of the House was opposed to the Commission. Within a few hours of the publication of the Commission's recommendations the Motion had attracted over 160 signatures and eventually 219 backbenchers signed.[74] As the signatures were being collected, WARA sent out letters to individual MPs who had not yet signed the Motion asking them to give it their support. The main difficulty facing the All-Party Committee was that it was not easy to maintain a truly all-party following. For example Conservative signatures to the Motion eventually outnumbered Labour by three to one. Part of the explanation of this was that nearly all the constituencies likely to be affected happened to be represented by Conservatives, though a number of potential Labour supporters refused to sign as a result of the bitterness among backbenchers which had been created by the Industrial Relations Bill. Nevertheless WARA, which was determined to handle the Parliamentary aspect of the P-R campaign themselves, made considerable efforts to encourage more Labour support for the Motion. As early as September 1970 (i.e. long before the Motion was published), Fennell had written to Sir Geoffrey de Freitas (Lab., Kettering), asking whether he could suggest any Labour Members who might sympathize with the cause and in January 1971, Fennell personally wrote to several Labour MPs whose names had been suggested by Niall MacDermot.

Finally, having achieved a very considerable impact with the Early Day Motion (especially amongst Conservative backbenchers, two-thirds of whom signed) the All-Party Committee went on to 'stage manage' the ensuing debates on Roskill in the two Houses. The Committee was again helped by WARA who arranged for Burson-Marsteller to circulate Peers and MPs with criticisms of the Roskill Report. Of course WARA's attempts to discredit Roskill's findings were helped immeasurably by Buchanan's description of the choice of an inland site as 'nothing less than an environmental disaster'.[75] In fact Buchanan's dissenting Report in many ways echoed WARA's own criticism of Roskill's findings and concluded by arguing that a Government decision 'which conceded the importance of the environment (as would be the case if Foulness was chosen, even allowing for the losses involved) would be an event of great significance for the future of Britain. It would show that the country . . . in spite of economic difficulties, is prepared to take a stand.'[76] Clearly Buchanan's considerable status as a planner not only gave weight to WARA's own criticism, but also cast authoritative and independent doubt upon the whole basis of Roskill's conclusions. Apart from sending a detailed brief to MPs, WARA also stimulated a large write-in campaign which was very nearly counter-productive. At one stage MPs began to receive so much anti-airport literature and correspondence that there was a serious risk of 'overkill'. Fortunately for WARA, Benyon (who had acted as Parliamentary 'postman' during the postal strike) managed to convince fellow MPs that the excessive zeal of WARA supporters was merely a manifestation of the deep-rooted hostility to the airport felt by the 'Buckinghamshire peasantry'.[77]

Assisted by the groundswell of public and Parliamentary opinion stimulated by WARA, the All-Party Committee contacted the Speaker of the House (who had himself been a Vice-President of WARA before his appointment as Speaker) indicating that a large number of backbenchers wished to speak in the debate on the Roskill Report and asking that frontbenchers might be urged to be brief. Similarly the Committee urged its supporters to limit their speeches to ten minutes to enable the Speaker to call as many backbenchers as possible. In the event, when the House of Commons debated the Report on 4 March 1971, no less than half of the forty-two speakers in the debate had previously signed the Early Day Motion. Thus the All-Party Committee had, in close liaison with WARA (and to a certain extent with other airport groups), managed to demonstrate very strong backbench opposition to an inland site. On the day of the debate, in which Mr Noble (the Minister for Trade) announced that the Government had not yet made up its mind and would 'listen to the voices',[78] WARA issued a statement answering the charges that it had waged an unfair high-pressure campaign. Fennell argued that there had been no high-powered publicity campaign but that the plan to site the huge airport on an inland site had brought forth an instantaneous outcry directed at the only people who could make the decision – the people's elected representatives at Westminster.

WARA had gained a good deal of publicity for its cause and this had increased once the Roskill findings had been announced. The Association missed little opportunity for publicizing its case although, as we have argued, it is wrong to attribute all this to professional P-R techniques.

The All-Party Committee's success in the Commons had been preceded by an even greater show of anti-Roskill feeling in the Lords' debate in February 1971. The debate was on Lord Molson's Motion taking note of the Roskill Report and, in winding up the debate, Molson observed that he could recall no other Commission which had made a recommendation that had subsequently been debated 'and not a single speech has been unequivocally in support of the recommendations made'.[79] It was against this background of fierce Parliamentary hostility that the Government had to formulate its decision.

**A decision is reached**

Almost immediately after the publication of the short version of the Roskill Report on 18 December 1970, press speculation began on the possibility that the Government would reject the Commission's findings and would choose Foulness. For example the air correspondent of *The Times* predicted, on the day after publication, that the Government would turn to Foulness if the Cublington protest became too strong. If the outburst was as large as that which led to the abandonment of Stansted, he argued, then there would be strong incentive to look to the more expensive coastal site.[80] This theme was taken up by almost every national newspaper, and of course encouraged WARA to increase its efforts to impress upon the Government the strength of feeling against the airport. They were given every encouragement by their contacts in government circles. A Cabinet Minister wrote to a member of the WARA Executive shortly before Christmas agreeing that there should be vigorous resistance, and stating that he was absolutely of the opinion that there was no question of the airport going to Cublington. He was very optimistic and predicted that if it did happen then it would be over his corpse.[81] There was every reason to believe that a significant proportion of the Cabinet shared this view from an early stage – all the more so because the Government had with great publicity created the new Department of the Environment. Thus, as with the heavy-lorries controversy discussed in chapter 7, the Government was quickly faced with a test of its sincerity towards the environment. The Roskill Commission added to this difficulty as it openly conceded that on planning and environmental grounds, Foulness was preferable to Cublington or to Thurleigh.[82] It had selected Cublington basically on grounds of construction costs and travel time. The Government's difficulty was compounded by the fact that a widespread distrust of Roskill's methods of inquiry and evaluation had developed. The exercise was a novelty in terms of policy-making in Britain and it found many informed critics. For example, the Town and Country Planning Association published a statement, in February 1971, accusing the Commission of having been led by its own methods of analysis into being unduly credulous where airport accessibility was concerned, and unduly sceptical over environmental factors. A similar view had earlier been expressed by Professor Peter Self in an article entitled, 'Roskill: nonsense on stilts'; he argued that greater rationality was not helped, but hindered, by the use of notional monetary figures which either concealed relevant policy judgments or involved unrealistic and artificial degrees of precision.[83] This view was undoubtedly shared by many MPs, and by certain members of the Cabinet, and so WARA's protest fell on highly fertile ground at Westminster. If, as was rumoured, the Cabinet was deeply divided on the issue, then the massive Parliamentary opposition engineered by the All-Party Committee and WARA, together with the heavy media coverage of WARA's efforts in the closing months of the campaign, managed to sway opinion in WARA's favour. There was little real surprise when, on 28 April 1971, John Davies (the Secretary of State for Trade and Industry) announced to the House of Commons that Foulness was to be the site of London's third airport. The Government believed, he said, that the irreversible damage that would be done to large tracts of countryside and to many settled communities by the creation of an airport at any of the three inland sites studied by the Commission was so great that it was worth paying the price involved in building the airport at Foulness.[84]

In effect the Government had taken the line of least resistance for there was only token opposition to the Foulness site and yet almost universal opposition to any inland site. Also the climate of opinion was such that 'the environment' had become a central political issue, and at relatively little extra cost to the taxpayer the Government, in selecting Foulness, was able to claim that – to use the Prime Minister's words – 'for the first time a government taking a major national decision has given pride of place to the protection of the environment'.[85] Whether or not such self-praise is truly deserved is open to debate. Derrick Wood, the Chairman of the then Action Committee Against Foulness Airport, would no doubt strongly disagree. Looking back on the airport controversy in a programme on BBC television, he has suggested that Foulness suffered the disadvantage of having no newspaper proprietors, no interested bankers and no leading politicians living in the area and that, far from gimmicks being the cause of WARA's success, it was their 'deep-seated influence' which won the day. This is in marked contrast to the earlier statement made by Peter Walker that the decision to select Foulness had nothing to do with the efforts of the Wing Airport Resistance Association. Perhaps the most accurate analysis has been provided by Bernard Levin when, shortly before the final decision was announced, he argued that the uproar over Stansted showed that any inland site within reach of London was a political impossibility. What followed, including the Roskill Commission, was nothing more than an empty ritual.[86]

## Notes

1 We should like to thank a number of people who gave us invaluable help in preparing this account. Our Research Assistant, S. K. Brookes, helped to sift and collate the documents. W. Benyon, MP, and S. Hastings, MP, helped us with the Parliamentary aspects of the campaign; the Editor and News Editor of *The Bucks Herald* gave us access to their files; J. Pargeter and Mrs G. E. Hetherington, who ran the WARA office, helped us to find our way through the documents; Lady Hartwell, E. R. A. de Rothschild, and Claude C. J. Simmonds also helped us on various points of detail. We are especially grateful to Mr and Mrs Desmond Fennell for their patience with our questioning, and for reading an earlier draft of this chapter.

Finally we should like to record our indebtedness to the WARA Executive Committee for agreeing to make their voluminous documents available for research, and it is to be hoped that future campaigning organizations will follow their excellent example in this respect.

2 H.C. 233, 1960-1.

3 Quoted in *Commission on the Third London Airport, Report*, HMSO, 1971, p. 3 (known colloquially as the 'Roskill Report' after Mr Justice Roskill who chaired the Commission).

4 *The Third London Airport*, Cmnd 3259, May 1967.

5 B. Cashinella and K. Thompson, *Permission to Land*, Arlington Press, 1971, pp. 34-5.

6 *Hansard*, 29 June 1967, cols 7-8.

7 *Daily Telegraph*, 4 June 1967.

8 Cashinella and Thompson, op. cit., p. 30.

9 For a full description of the campaign see O. Cook, *The Stansted Affair*, Pan, 1967.

10 *Hansard*, 22 February 1968, cols 667-9.

11 B. Levin, 'The empty ritual of Roskill', *The Times*, 30 March 1971.

12 A. Crosland, 'The folly of Foulness', *Guardian*, 28 April 1971. Also accepting the rejection of the inland sites, he argued that technology might provide a solution through the development of aircraft capable of short or vertical take-off.

13 *Hansard*, 22 February 1968, col. 671.

14 For the full terms of reference, see *Hansard*, 20 May 1968, cols 38-40.

15 Roskill Report, p. 22.

16 Letter, 7 November 1968.

17 Letter, 8 December 1968.

18 Letter, 12 November 1968.

19 Letter, 20 February 1969.

20 Letter, 25 February 1969.

21 A certain amount of confusion has reigned over the naming of the site. The existing aerodrome was known to the local population as 'Wing', though Roskill mostly uses 'Cublington'. The two villages are about three miles apart.
22 Roskill Report. Papers and Proceedings, vol. VIII, part 2, section 4, p. 3. The summary of findings presented in this section derives from this report.
23 Letter, 6 March 1970.
24 Cublington figures from the Essex survey report (Roskill), op. cit., p. 24; national figures from D. E. Butler and A. King, *The British General Election of* 1966, Macmillan, 1966, p. 264.
25 WARA Constitution.
26 Executive Minutes, 20 April 1969. For a more detailed account of this and other incidents, see D. Perman, *Cublington. A Blueprint for Resistance*, Bodley Head, 1973, pp. 88-92.
27 *Buckingham Advertiser*, 7 February 1969.
28 Interview with Fennell, 5 August 1972.
29 *Buckingham Advertiser*, 18 April 1969. Lord Roseberry later publicly stated that he was in no way connected with WARA.
30 Executive Minutes, 27 April 1969.
31 For a discussion of the classical expositions of this point see G. Parry, *Political Elites* (ch. 2), Allen & Unwin, 1969.
32 Executive Minutes, 30 June 1969.
33 The Noise and Number Index (NNI) is a composite variable which takes into account the noise level and the number of aircraft per day. This has been shown to correlate highly with the degree of annoyance people experience with aircraft. For fuller details see *The Committee on the Problem of Noise* (*The Wilson Committee*) *Final Report*, July 1963, Cmnd 2056.
34 Letter, 30 June 1970.
35 See *Wing Airport Urbanisation Proposals* and *Wing . . . and the Way We Live*, two pamphlets published by WARA.
36 'We'll use force to save Cublington', *Observer*, 20 December 1970.
37 Letter, 19 May 1969.
38 *The Times*, 5 April 1971.
39 Executive Minutes, 30 June 1969.
40 Letter, 20 May 1969.
41 Letter, 5 May 1971.
42 Letter, 14 May 1970.
43 Letter, 2 April 1970.
44 Letter, 23 October 1969.
45 Letter, 14 April 1970.
46 WARA press release.
47 Letter, 24 September 1970.
48 Executive Minutes, 28 June 1970.
49 Letter, 8 December 1970.
50 Letter, March 1970.
51 CPRE press release, 18 December 1970.
52 CPRE press release, 14 January 1971.
53 Notes of R. B. Verney (Chairman of Buckinghamshire County Planning Committee) for a meeting on 28 March 1969.
54 Letter, 24 April 1969.
55 Letter, July 1969.
56 Executive Minutes, 27 April 1969.
57 See, for example, Executive Minutes, 20 April and 4 June 1969, and letter, 26 August 1969.

58 For a more extensive description of this and other activities organized by the local branches of WARA see D. Perman, op. cit.
59 WARA Information Sheet, no. 11.
60 Cashinella and Thompson, op. cit., p. 10.
61 Ibid., p. 14, our emphasis.
62 *Campaign*, 30 April 1971.
63 Executive Minutes, 5 April 1970.
64 Executive Minutes, 6 December 1970.
65 Letter, 13 May 1971.
66 S. E. Finer, *Anonymous Empire*, Pall Mall, 1966, p. 56.
67 Ely Devons, 'Government and the inner circle', *Listener*, 27 March 1958.
68 R. Rose, *Politics in England*, Faber & Faber, 1965, p. 180.
69 For an interesting discussion of this point, see T. Lupton, 'The social background and connections of top decision-makers', *Manchester School*, 1959.
70 *Hansard*, 30 November 1970, cols 879-80.
71 Burson-Marsteller, Contact Report, 3 July 1970.
72 Letter, 13 August 1969. (For a discussion of all-party committees see J. J. Richardson and Richard Kimber, 'The role of all-party committees in the House of Commons', *Parliamentary Affairs*, Autumn 1972.)
73 These are motions tabled by backbench MPs which appear on the Order Paper for discussion on 'an Early Day'. An Early Day seldom arrives, however, and the motions remain undebated expressions of backbench opinion. See S. E. Finer *et al.*, *Backbench Opinion in the House of Commons*, Pergamon Press, 1961.
74 Including three Members who signed twice.
75 Roskill Report, p. 149.
76 Ibid., p. 160.
77 Letter to *The Times*, 14 April 1971.
78 *Hansard*, 4 March 1971, col. 1913.
79 *House of Lords Debates*, 23 February 1971, col. 1051.
80 *The Times*, 19 December 1970.
81 Letter, 2 December 1970.
82 See Roskill Report, paras 13.21, 13.60 and 14.12.
83 *Political Quarterly*, vol. 41, no. 3, July/September 1970, p. 260. See also E. J. Mishan, 'What's wrong with Roskill?', *Journal of Transport Economics and Policy*, September 1970.
84 *Hansard*, 26 April 1971, col. 35.
85 *The Times*, 28 October 1971.
86 *The Times*, 30 March 1971.

# Chapter 9

# Conclusion: tactics and strategies

The only sure defence against having some technological monstrosity in your neighbourhood is to have some technological monstrosity there already which is inimical to the proposed new one.

'Way of the world', *Daily Telegraph*, 24 February 1972

## I

As Dr Roy Gregory points out in his book *The Price of Amenity*, there is a good deal to be said for allowing case-studies to speak for themselves. Certainly, the case-studies presented in this book are sufficient neither in number nor in scope to form the basis for rigorous generalization. They do offer the opportunity, however, for making some more limited observations on the circumstances in which environmental pressure groups in Britain can hope for success. The factors influencing the success, or failure, of environmental groups have been organized under seven general headings though these are neither definitive nor mutually exclusive, but are empirically useful. These are: 1 Advance intelligence, 2 Liaison with administrators, 3 Rational argument and the merits of the case, 4 Relationships with legislators, 5 Relationships with the mass media, 6 Resources, 7 Sanctions.

### 1 *Advance intelligence*

In describing the work of the Wing Airport Resistance Association (chapter 8), we emphasized the effort which was put into sounding out the opinions of experienced politicians and administrators about the procedure which the Government was likely to adopt in dealing with the Roskill Report. We also noted that Professor Finer has characterized this activity as a necessary condition for success. From WARA's point of view, the way in which the Report was to be handled was crucial. A government decision announced simultaneously with the publication of the Report (in the style of the Labour Government's initial decision to select Stansted) would have demanded rather different tactics from those which were actually adopted. As it was, the Government's attitude allowed, indeed almost encouraged, the resistance associations to organize mass opposition to the Roskill recommendations.

Of course, gathering advance intelligence is not confined to such matters of procedure. Groups must also learn to tell in which direction the 'policy wind' is blowing. This can be done in three ways, by establishing a good working relationship with decision-makers in firms and institutions, by monitoring their public statements thoroughly, or by what amounts to environmental espionage. In many instances groups are dependent upon individuals coming forward with information, in the way that some midland lorry drivers revealed details of illicit cyanide tipping to members of the Conservation Society. The press also plays a useful part in alerting groups, and sometimes obtains information which is not intended for general consumption. For example in October 1972 the *Sunday Times* published a set of secret and supposedly tentative proposals for a drastic reduction in the overall size of Britain's railway network. This immediately provoked a hostile reaction from a number of groups concerned with the broad environmental aspects of the transport problem, and from narrower sectional interests. The Minister for Transport Industries protested, predictably enough, that the proposals were but one of a whole range of possible options; yet, whatever the final outcome of the deliberations within the Department, it is clear that, by gaining advance warning of departmental thinking, outside groups have been placed in a potentially stronger position. No doubt this explains why the much promised 'open government' has failed to materialize, and probably never will. The more usual situation, of environmental groups being presented with decisions, or at most a restricted range of options, and having to fight a rearguard action to reverse or alter them, is likely to persist.

The need for reliable advance intelligence is even more apparent at the local level, where local civic groups are constantly fighting a rearguard action to reverse the planning decisions of local authorities. The now notorious case of Harrogate, where controversial plans were submitted to the planning committee only four days before the meeting, is a sad reminder of the predicament of local civic groups faced with a determined planning authority. In response to criticism from the Secretary of the Harrogate Society, the Deputy Town Clerk is reported as saying that it was not possible to inform everyone of the planning applications received, but that 'if objectors had inspected the Register during the four days preceding the committee meeting in question they would have seen that a new application had been made'.[1] This followed earlier accusations made by the Society that important planning decisions were being taken in secret, which was 'a deliberate attempt to prevent public discussion of the Council's controversial policies'.[2] Because of the difficulty of persuading policy-makers, even at the local level, to reverse policy decisions once they have been announced publicly, local civic groups must regularly scrutinize local planning registers if they are to stand any chance of mounting an early campaign. The price of amenity is eternal vigilance and, as S. E. Finer says, vigilance is a full-time job.[3] In his study of the siting of the M4, Dr Gregory describes how local people noticed surveyors at work in their area before the proposals were publicly announced, and the first intimation some people had of the intentions of the mining companies towards Snowdonia was the appearance of these same harbingers of environmental destruction. In chapter 4, H. R. Burroughes clearly illustrates the consequences for amenity groups if they fail to mount an effective campaign until after firm proposals have emerged. Although governments and other public bodies do sometimes back down (as happened over the Stansted decision) this can only be expected to happen in special circumstances; for example, when highly influential people are threatened by the decision, when a government is particularly anxious not to alienate public opinion, when the course of events causes a government to rethink its policies and their implications, and so on.

## 2 *Liaison with administrators*

The battle fought by the amenity movement against the manufacturers' proposals to introduce heavier lorries on to Britain's roads highlights the advantages to environmental groups of developing a working relationship with civil servants: initially the Construction and Use Regulations appeared to fall within the normal departmental consultations with the manufacturers on purely technical details. With the growth of environmental awareness, problems which hitherto have been regarded as 'technical and detailed' may gradually be seen as having much wider environmental consequences and amenity groups should seek to be included in the sounding-out process which is so important in British Government. Some progress has been made; the Faversham Society succeeded in widening the consultative network during the 'heavy lorries' campaign and the Conservation Society and other groups produced memoranda during the preparatory period before the Stockholm Conference.

Undoubtedly there is a vast range of problems, often apparently of minor technical significance and raising no overtly political difficulties, which are currently settled by normal consultations between civil servants and the affected interests. One important task of groups concerned with the wider environmental brief is to impress upon government departments the increasingly obvious point that 'the environment', by even a fairly narrow definition, is as much an affected interest as are the sectional interests which have long been part of the consultative network. As yet, the majority of groups concerned with environmental problems, including groups of long standing, are not *automatically* consulted on matters which legitimately concern them. Thus, rather surprisingly, the CPRE is reported as asking the Secretary of State for the Environment if it could be allowed to submit evidence to the committee appointed to consider planning control over mining operations.[4]

Chapter 3 suggests that a group which can develop a reputation for being moderate, responsible and expert in its field, as the National Society for Clean Air has done, stands a greater chance of becoming an established part of a department's 'clientele'. By having two of its members appointed to the Beaver Committee, the Society was even able to influence the policy proposals which were presented to the Government. Although militant and unorthodox approaches may be appropriate in some circumstances, a group which persists with such behaviour is unlikely to be accepted as part of the regular consultative machinery. Of course, even when consultation takes place, considerable persistence may be required in the face of skilled administrative 'stonewalling'. Wayland Kennet has given an unusually detailed account of the kind of resistance which even ministers may receive from civil servants.[5]

## 3 *Rational argument and the merits of the case*

Most development issues involve a fine balance of economic against environmental benefits and it is one of the tasks of amenity groups to present policy-makers with a balance of argument which is weighted less in the favour of economic interests than has hitherto been the case. Meeting this challenge is probably the most difficult task facing environmental groups, simply because they lack the necessary resources to finance adequate research. Also, there is the further problem of choosing which battles to fight. Groups must be selective, partly because they have limited resources, and partly because some instances of environmental degradation will, on balance, be in the public interest. Outright opposition to everything regardless of the merits of the case is not only an irrational policy, but is also likely to prove counter-productive.

However, a mistake that is frequently made is to suppose that difficult environmental problems can be satisfactorily resolved solely by 'experts'; that all that has to be done is to appoint a body of experts to recommend solutions based on sound technical arguments. The Roskill Commission was certainly regarded in this light by the Government. The same might also be said of the Zuckerman Commission which investigated the problem of mining in the National Parks on behalf of the mining industry. The essential point is that solutions which are technically desirable may often be politically unacceptable.

Despite the view that the Roskill Commission was an empty ritual, WARA had no alternative at the time but to present a properly reasoned case against the selection of Cublington and the other inland sites. The technical case it presented, particularly the evaluation of the cost-benefit analysis, at least impressed many MPs and succeeded in sowing seeds of doubt in their minds about the approach which the Commission was using. Similarly, in the M4 controversy, Dr Gregory concluded that any objector who wishes to be taken seriously has to restrict himself to arguments and data that may be genuinely useful in reaching a decision. As a corollary to this he points out that, by forcing a ministry (or a local authority) to go through the exercise of looking again, an objector obliges it to check its facts and figures and to reconsider the logic of its case. If the objector loses, at least he has lost to a scheme which is unlikely to have hidden defects, and he may have been able to obtain some concessions, as in the case of Manchester's water scheme.

Above all, amenity groups must be able to present a viable alternative to the proposed scheme. This was the greatest weakness both of the opponents of Manchester's water scheme and also of the protesters against the high-voltage transmission line across the South Downs. In the former case the experts were unable to fault Manchester's proposals and therefore could not provide a satisfactory alternative. In the latter case, the only viable alternative was to place the cable underground, the cost of which would have been astronomical.

That the technical expertise which environmental groups often need is very expensive can be seen from WARA's balance sheet. Few groups can hope to raise, year after year, even a fraction of the sum spent by WARA. Thus the main hope, particularly of local groups with limited funds, is to tap as much free technical advice as possible, and to utilize any personal contacts that may be available to obtain it. Even if substantial funds exist, groups may need to use such personal contacts in the scramble that develops for the best experts, once battle is engaged. Indeed there is the further problem that there are some spheres in which the supply of independent experts is strictly limited; for example there are few experts on electricity transmission outside the electricity industry.

Occasionally, environmental groups may find support in unexpected quarters. In chapter 7, for example, we saw that the local authority associations and the Government's own Road Research Laboratory produced cogent economic arguments opposed to the manufacturers' submissions. On the other hand, groups are sometimes unable to get the information they need to compile their case. Monitoring pollution is not an easy task for a voluntary association and this problem is magnified by the laws of secrecy relating to manufacturing processes. Under these, Pollution Officers or other local officials are unable to publish analyses of industrial discharges. This secrecy might be more acceptable if there were any grounds for having confidence in the operation of regulatory bodies like the Alkali Inspectorate.[6] The Government has however indicated that future legislation will allow the public to have more information than at present.[7]

Finally, it must also be pointed out that a group's success may be influenced by chance factors. There is no doubt that the smog of 1952, which received massive press coverage, was of considerable help to the National Smoke Abatement Society in pressing home its case.

## 4 *Relationships with legislators*

Several of the case-studies in this book point to the importance of having a good working relationship with both Parliament and local authorities. In the event, WARA's key battle was the parliamentary one; were it not for the activities of the All-Party Committee, ably supported by WARA, it would have been much easier for the Government to have selected Cublington had it wanted to do so. Despite the widespread view that Parliament is in decline, there are occasions on which substantial parliamentary backing can prove a considerable asset to a pressure group. This is especially so when the group is not involved in the earlier stages of policy formation, and must therefore rely upon applying political pressure on the minister concerned. The capacity of backbenchers to develop a cross-bench, and therefore potentially influential, approach to complex political issues should not be underrated;[8] though because MPs have such limited facilities and indeed time, which can be devoted to acquiring information on a given issue, a group seeking support must expect to supply MPs with all the necessary information (as indeed WARA did). When a group is well established and has something to contribute, it may even be approached by MPs seeking help for a legislative project – just as Gerald Nabarro approached the National Smoke Abatement Society before attempting to initiate clean air legislation. Groups wishing to promote legislation should certainly encourage friendly MPs to enter the ballot for Private Members' Bills with a view to promoting their Bill, and they may even be able to persuade an MP who has made a successful, but speculative, bid for Private Members' time to promote their legislation.

In a more general way, lobbying MPs can help to create a climate of opinion at Westminster which is favourable to environmentally orientated policies. On the other hand, it must be recognized that, if lobbying becomes excessive and appears too well organized, it may prove counter-productive. In fact, WARA came dangerously close to alienating a section of Parliamentary opinion because of the sheer volume of protests delivered to MPs. In some circumstances the House of Lords may have an important role to play, as it did in the case of Manchester's water supply and during the post-Roskill phase of the airport controversy. As with the Commons, there is a considerable degree of sympathy in the Lords towards environmental matters, and individual Peers have shown willingness to espouse the amenity cause.

Many of the day-to-day decisions affecting the environment are taken at the local level and in some ways local councillors are more exposed to public pressure than national political figures. Certainly the attitudes of local authorities can be extremely important. Dr Palliser points out in chapter 2 that the fortunes of the York Civic Trust were greatly improved on the appointment of a committee chairman who was more favourably disposed towards ideas in the Esher Report. WARA, too, was greatly assisted by the fact that the local authorities concerned were strongly opposed to any inland site and recognized WARA as the mouthpiece for grass roots opinion in the area. This is in sharp contrast to the complex of disputes and wrangles between the local authorities, at various levels, involved in the electricity pylons dispute – conflict which must have strengthened the Generating Board's position.

Amenity groups should certainly seek to establish themselves as part of the formal consultative machinery at the local level. However, as Dr Palliser points out, many councillors see such moves (in the form, say, of advisory panels) as a challenge to their position as elected representatives, and local authorities are probably even less inclined to institute consultative processes than is the central government.

Finally, local authorities may be a source not only of political support but also may provide technical assistance, since they possess the research resources necessary for the preparation of an effective brief. Thus, for example, the Reading traffic survey produced new data on the likely effects of alternative motorway routes; and, in the campaign for clean air, local MOHs produced documentary evidence on the effects of heavy smog on health.

## 5 *Relationships with the mass media*

Political scientists have hitherto regarded using the mass media as something of a last resort for British pressure groups. There is some evidence, however, to suggest that, so far as environmental pressure groups are concerned, resort to the mass media can prove an early and effective precursor to more conventionally accepted forms of political influence. There is little doubt, for example, that the initiative taken by the *Sunday Times* in the 'heavy-lorries' controversy played an important part in creating a climate of opinion hostile to the motor manufacturers' proposals. The newspaper coverage of the issue laid the basic groundwork on which the Civic Trust was later able to build with detailed documentation. In a similar way the National Smoke Abatement Society propagated the concept of 'clean air zones' largely through judicious use of newspaper and television coverage; and the space which the media devoted to the problem of cyanide dumping at least speeded up Government activity even if it did not precipitate it.

Since media coverage of environmental problems has increased during the last few years, with many newspapers having environmental correspondents, it is relatively easy for groups to arrange for environmental issues to have an airing. This is particularly so given the strong commitment of many journalists and broadcasters to the cause of environmental protection; indeed, some of them have formed an 'Environmental Communicators Organisation' (ECO) which is 'dedicated to promoting a constant awareness of the forces which threaten human survival'.[9] ECO believes that the environmental crisis should be as newsworthy as are unemployment and war. From the point of view of environmental pressure groups the favourable attitudes of 'communicators' are likely to prove a valuable asset, sometimes forcing central government, industry, and local authorities alike on to the defensive and forcing them to respond, in the face of 'public opinion', to the demands of the environmental lobby.

## 6 *Resources*

Perhaps the most obvious necessary condition for success is that groups must have sufficient resources for the campaigns they undertake. This entails not only the capacity to raise funds, but also the existence of a pool of technical and organizational ability, and leadership. Both chapters 5 and 8 illustrate the importance of good leadership, though it is usually easier to find the leaders for the kind of *ad hoc* battles described in those chapters than for the more continuous activities undertaken by many local civic societies. Leadership, of course, requires the support of an active membership; it is a moot point, at present, what proportion of the many local amenity societies consists of two or three active individuals and of no one else. In many situations a group's influence is dependent upon its ability to demonstrate that it represents a significant section of the community; WARA rightly perceived that it was in such a situation and produced over 60,000 signatures in support. Local amenity groups of a more permanent nature should make an effort to maximize their membership and should try to involve as many as possible in the work of the society. This is somewhat easier in those communities which have a close-knit social network.

Another resource, which WARA found invaluable, is that of access to key figures in Westminster, Whitehall or some other part of the political system. If, on any particular occasion, they are unable actually to influence the course of events, such individuals may still provide useful advance intelligence, or can explain the procedures that are normally used in arriving at decisions, and may have valuable advice to offer.

Although many people find administrative chores exceedingly dull and unrewarding, it is important that they be performed efficiently within a sound organizational structure. WARA's organizational difficulties were compounded by the fact that it had to divide its energies between technical and political opposition, but the key to WARA's success in coping with this was the manner in which the routine administrative tasks were performed, using the same approach to keeping and filing records and dealing with correspondence which is essential to any viable business enterprise. Those groups which are run by people who despise formal organization and efficient administration for ideological, temperamental or other reasons are likely to find themselves in a strategically weak position, and in the long run these groups are unlikely to persist.

Important though these various resources are, it is the resource of finance which so often determines what a group may achieve. The extremely high cost of mounting a good technical argument against development schemes had led some people to argue for state aid to amenity societies. Indeed, the Government's own Stevenson Committee recommended that:[10]

> In the case of groups involved in planning procedures, the DOE should commit itself to evolving means whereby financial assistance can be given to voluntary groups where it is necessary to enable a point of view to be properly presented or an issue to be fully investigated.

This recommendation was subsequently supported by Wayland Kennet (Parliamentary Secretary responsible for the environment in the 1966-70 Labour Government), who argued for its implementation in a letter to *The Times*.[11] Clearly what is needed is a technical counterpart to legal aid.

## 7 *Sanctions*

A pressure group is in a strategically strong position if it has a sanction which it can apply. At first sight, environmental groups may seem to stand little chance of bringing effective sanctions to bear on policy-makers but in fact there are two weapons which groups may use. The first is common to all pressure groups, though it may be somewhat ephemeral: that of gaining, or appearing to gain, a sufficient weight of public opinion behind a set of proposals.

We have already referred to the interest the media have in environmental matters. Whether this is the cause or effect of a more general awareness of environmental problems is uncertain, but it is clear that at the moment many groups can generate a body of opinion to which policy-makers are prepared to respond. The outcry over 'heavy lorries', over cyanide dumping, and over the demolition of particular historic buildings has clearly demonstrated this.

The second weapon which may be used is more specific to environmental pressure groups. In chapter 6, Dr Gregory has described the procedural stages associated with planning the M4. It does not take lateral thinking to realize that such a procedure offers considerable possibility for those bent on delaying any proposed development. In short, the opportunity to put one's case against a proposal may be used not only for just that, but also as a tactic. For example, in June 1969, one of the more active branches of the CPRE, the Sheffield and Peak District Branch, threatened to oppose the proposed expansion of a cement works in the Hope Valley and thereby cause a very considerable delay to the company concerned. By making this threat the branch managed to gain important concessions from the company, which enabled the Peak Planning Board to impose conditions on the planning permission that the company would otherwise have contested.[12]

Thus the sheer complexity of statutory procedures involving the resolution of planning disputes, while being in some ways a costly burden to amenity groups, may also be turned to their advantage. No rational authority or company will willingly provoke amenity protests if the procedure for reaching a decision is as time-consuming as that for the line of the M4. Even so, amenity bodies should note that other, less environmentally conscious, concerns have also spotted the advantages of exploiting existing procedures. The House-Builders' Federation has recently encouraged its members to make more use of the appeals machinery in an attempt to take planning decisions away from local authorities. Consequently, the number of appeals increased by 50 per cent in May 1972 (compared with May in the previous year) at a time when the Minister had asked his inspectors to take a more positive attitude towards the need for land release when considering appeals.[13]

The very difficulty of reaching a final decision may itself prompt development authorities to search for less damaging means of achieving their goals. It is reported, for example, that the CEGB is considering the possibility of building generating complexes of 10,000 megawatts each, five times the size of the biggest power stations in Britain today, 'as a solution to the growing public resistance to an expanding network of smaller power stations'.[14] The fact that modern corporations are increasingly anxious to present a good public 'image' can only assist environmental groups in applying successful pressure against development proposals.

Exploitation of the planning procedure must, like all weapons, be used sparingly, for there is a danger that Government will respond to undue delays by revising the planning procedure. Indeed, there are signs that this may already be happening. The Town and Country Planning Amendment Bill eliminated the right of every objector to be heard at public inquiries into overall structure plans, such as the Greater London Development Plan. Under the new arrangements, *selected objectors* are invited by the DOE to take part in a less formal public hearing of the key issues, which the DOE also selects. As Judy Hillman points out, amenity groups may not be impressed by the argument that the Minister may leave particularly difficult problems until the local plan is drawn up, thereby giving them full opportunity to object.[15] He may not do this. Whether or not groups are disadvantaged will become clearer as we see the new provisions in operation.

## II

The increasing awareness of environmental problems has had a marked effect on the pattern of groups concerned about these problems. Not only have long-established groups witnessed a sudden increase in membership (e.g. the membership of the National Trust at the end of 1971 stood nearly 37 per cent higher than twelve months previously), but also they have seen a proliferation of new groups, often having a much wider concern for the environment. The Conservation Society, for example, was founded in 1966 and while paying due attention to aspects of conservation concerning the countryside, wildlife, historic buildings and so on, believes that 'piecemeal measures are not enough'.[16] It sees the non-stop growth of population and material demands as the fundamental threat to the environment and thus concentrates on the need for a stable population policy and zero economic growth. Thus they see a choice between 'a policy which regulates our numbers and economic behaviour, and catastrophe'.[17] The proliferation of groups has, of course, emphasized the problems of co-ordination and co-operation among groups. In chapter 7 we saw how CoEnCo failed in its co-ordinating role, leaving the various interested bodies to pursue individual campaigns against the manufacturers. In this case the lack of co-ordination had no serious consequences for the amenity movement, but so long as groups have very limited funds duplication of work is likely to weaken their overall impact.

The multiplication of groups representing environmental interests and their posture of independence from one another is in sharp contrast to the trend among more conventional sectional groups. The amenity movement has not yet shown signs of evolving a structure comparable to the CBI or the TUC. This is not to suggest that the movement is unaware of the weaknesses of unco-ordinated campaigning; the Conservation Society itself chose the co-ordination of the work of other societies as one of its original aims. However, this move apparently caused a 'general raising of hackles' among existing amenity societies, and the offending words were later deleted from the Society's aims.[18] Since then, relations between the various groups concerned with amenity have undoubtedly improved, but in January 1972 the Director of the Conservation Society was hinting that sooner or later organizations like the CPRE would have to move from a position of concentration on specific, easily identifiable, amenity problems to face the complex and divisive questions of 'choosing between one good thing for one set of people and another good thing for another group, or how to weigh our present enjoyment against the consequences for our descendents'.[19] As he suggests, the conservation movement is likely to fall into disarray unless clearly defined positions emerge. Also, as John Barr suggested, if an all-party committee in Parliament is to command attention, it will need to be backed by a united voluntary amenity lobby.[20]

*Conclusion: tactics and strategies*

The transition from the recognition of the need for unity to its achievement presents virtually insuperable difficulties, simply because of the enormous range of interests and attitudes encompassed by the conservation movement. While the many local and national groups are united in the broad desire to maintain or improve the quality of the environment, they differ widely on the urgency of the predicament, the solutions advocated, and on the means to be used to obtain these solutions. Thus, as has been implied, the CPRE places an emphasis on using the statutory planning machinery to fight specific development proposals and often does so with considerable success, while other groups are committed to a more radical view, sometimes involving direct action. It is not easy to accept that the Oxford Street Action Committee, the Dwarves, the Street Farmers, and groups firmly embedded in the hippy world are part and parcel of a movement which also includes bodies like the Civic Trust, the CPRE, and the National Trust. It is unlikely that such a wide range of groups will find much common ground, particularly since many of the more radical groups see a fundamental change in the nature of British society as a prerequisite of any real improvement in the environment.

Just as it is unlikely that environmental groups will evolve a viable 'umbrella' organization analogous say to the CBI, so it is also unlikely that they will spawn a successful political party. This was advocated in the issue of the *Ecologist* containing the much discussed 'Blueprint for survival'.[21] What the proponents of an environmental political party tend to overlook is the considerable adaptive capacity of the existing parties. If environmental issues are seen to have a significant electoral impact, the existing parties are bound to make a sufficient response to prevent their supporters from going over to the new party.

More serious than any lack of organizational unity is the fact that the conservation movement is even divided over the seriousness of the threat to the environment. In such a situation it is unlikely that a government will respond with a coherent set of environmental policies, though it may propose worthwhile, but disparate, policies which may not go very far towards solving the fundamental problems. That governments do find the task of evaluating environmental arguments a particularly difficult one is shown by the appointment of the Roskill Commission, at unprecedented expense and with great ceremony, and the subsequent handling of the Report. After that exercise, it is doubtful whether politicians will be inclined to trust cost-benefit analyses of any but the most technical and obviously quantifiable situations. Similarly, the argument of economic benefits versus amenity in the National Parks cannot be resolved by expert commissions, whoever appoints them. The problems have to be faced by politicians and resolved by political decision, preferably in the context of a coherent approach to the environment. Because many of the battles fought by conservationists raise fundamental social issues and challenge many existing cultural values, the campaign for the environment is bound in the end to be a political one, and may possibly even test the capacity of the political system itself.

## Notes

1 See *Guardian*, 11 August 1972.
2 *The Times*, 5 January 1972.
3 S. E. Finer, *Anonymous Empire*, Pall Mall, 1966, p. 56.
4 *The Times*, 12 October 1972.
5 Wayland Kennet, *Preservation*, Temple Smith, 1972, pp. 79-82.
6 For a critical discussion of this little-known body see J. Bugler, *Polluting Britain*, Penguin, 1972, ch. 1.
7 *Hansard*, 17 March 1972, col. 917.
8 See Finer, op. cit., p. 75.
9 ECO recruiting leaflet.
10 *Fifty Million Volunteers*, HMSO, 1972, p. 68.
11 *The Times*, 13 September 1972.
12 Civic Trust *Newsletter*, March 1972, p. 3.
13 *The Times Business News*, 13 October, 1972.
14 *The Times Business News*, 15 June 1972.
15 *Guardian*, 5 November 1971.
16 Pamphlet, *The Conservation Society*.
17 Ibid.
18 *Conservation*, no. 39, January 1972, p. 1.
19 Ibid., p. 2.
20 *New Society*, 5 February 1970.
21 *Ecologist*, vol. 2, no. 1, January 1972, preface.

# Further reading

The literature on 'the environment' is large and proliferating rapidly. There is a particularly large field of books which is concerned to draw attention to the existence and nature of the 'environmental crisis', but since this is peripheral to our topic it has been excluded. We have selected a few of the more interesting and useful works relevant to the themes of this book and grouped them under four general headings – though some works overlap several categories.

## I The history of preservation and the preservation of history

Short historical accounts of the growth of conservation and preservation may be found in:

Max Nicholson, *The Environmental Revolution*, Penguin, 1972, chs 6, 7 and 8.
Wayland Kennet, *Preservation*, Temple Smith, 1972, especially chs 1 and 2.
John Harvey, *Conservation of Buildings*, A. & C. Black, 1972.

The Council for British Archaeology has also published several pamphlets on the preservation of history including:

*Historic Towns*, 1965.
*Historic Towns and the Planning Process*, 1961.
*The Erosion of History: Archaeology and Planning in Towns*, 1972.

## II Case-studies

A number of case-studies exist which touch on aspects of the environment, but the following deal more centrally with the topic:

Roy Gregory's *The Price of Amenity*, Macmillan, 1971, is an excellent, but expensive, collection of well-researched case-studies of five environmental decisions: the struggle over the use of part of North Oxfordshire for ironstone working; the siting of a power station at Holme Pierrepont, near Nottingham; the siting of a reservoir at Cow Green, Upper Teesdale; the siting of a North-Sea gas terminal at Bacton, Norfolk; the building of a new gas holder at Abingdon.

Amory Lovins and Philip Evans have produced a fine, but again expensive, collection of photographs and essays on Snowdonia and the problems facing National Parks in *Eryri, the Mountains of Longing*, Allen & Unwin, 1972.

In *Battle for the Environment*, Fontana, 1972, Tony Aldous (Environmental Correspondent for *The Times*) presents an informed discussion of a number of environmental problems, with the emphasis on those produced by motor vehicles.

In *Polluting Britain*, Penguin, 1972, Jeremy Bugler (Environmental Correspondent for the *Observer*) examines six case-studies of pollution: air pollution and the Alkali Inspectorate; the pollution of the Mersey; the problem of marine waste disposal; noise abatement; industrial excavations and the London Brick Company; problems facing the National Parks.

Wayland Kennet, formerly Parliamentary Secretary with special responsibility for the environment in the Ministry of Housing and Local Government, gives an insider's view of decision-making at the Ministry, and also of several case-studies in his book *Preservation*, mentioned in Section I.

V. Payne, 'The History of the Soil Association', MA thesis, University of Manchester, 1972, provides a detailed account of a long-established association concerned with one particular aspect of the environment.

## III Directories and guides to action

There are several directories of organizations and groups interested in environmental problems, but the two most comprehensive are:

*The Countryside Information Directory*, 1970, published by the Countryside Commission (1 Cambridge Gate, Regent's Park, London NW1 4JY), a guide to what organizations do and what services they provide, and an *Environmental Directory*, rev. ed. 1972, from the Civic Trust (17 Carlton House Terrace, London SW1Y 5AW).

Peter Gresswell's *Environment: An Alphabetical Handbook*, John Murray, 1971, is a good general reference work for almost anything concerning the environment, although as new legislation is introduced it will become outdated.

*Further reading*

The pamphlet produced by the Council for British Archaeology (8 St Andrew's Place, London NW1 4LB), *Public Inquiries: Presenting the Conservation Case*, 1971, is a useful guide to the basic principles important in presenting a case at a public inquiry. R. E. Wraith and G. B. Lamb, *Public Inquiries as an Instrument of Government*, Allen & Unwin, 1971, is the most comprehensive study of the public-inquiry system in Britain; some aspects of this are dealt with in R. E. Wraith, 'Planning inquiries and the public interest', *New Society*, 22 July 1971.

Antony Jay in *The Householder's Guide to Community Defence Against Bureaucratic Aggression*, Jonathan Cape, 1972, has produced an amusing but interesting guide to action against bureaucracy.

The *Consumers' Guide to the Protection of the Environment*, Pan/Ballantine, 1971, by Jonathan Holliman, is a practical guide to the consumer who is interested in preserving and improving the environment. Similarly, John Barr has produced an 'action guide for the UK' in *The Environmental Handbook*, Pan/Ballantine, 1971, which is both lively and useful.

## IV Miscellaneous

Any reading list concerned with the environment is soon likely to become dated. Readers wishing to keep abreast of developments should find the following sources useful:

*Planning Bulletin*, published weekly by the Town and Country Planning Association (17 Carlton House Terrace, London SW1Y 5AS), is a digest of news and comment, with a section on books and articles.

*Town and Country Planning*, published monthly by the Town and Country Planning Association, contains more detailed articles on current developments.

The *Newsletter*, produced monthly by the Civic Trust (17 Carlton House Terrace, London SW1Y 5AW), contains information on aspects of the environment of concern to the Trust.

Similarly, the Conservation Society (34 Bridge Street, Walton-on-Thames, Surrey, KT12 1AJ) and the CPRE (4 Hobart Place, London SW1W 0HY) publish regular bulletins of general interest.

*Consernus*, the newsletter of the NUS Conservation Project (3 Endsleigh Street, London WC1H 0DU), contains news and comment from student groups.

For a digest of European developments see *Nature*, the monthly newsletter of the Council of Europe (from the Countryside Commission, 1 Cambridge Gate, Regent's Park, London NW1 4JY).